T0330455

MANAGING AND ENGINEERING COMPLEX TECHNOLOGICAL SYSTEMS

MANAGING AND ENGINEERING COMPLEX TECHNOLOGICAL SYSTEMS

AVIGDOR ZONNENSHAIN AND SHUKI STAUBER
The Gordon Center for Systems Engineering, Technion, Haifa, Israel

Published by John Wiley & Sons, Inc., Hoboken, New Jersey
Published simultaneously in Canada

Library of Congress Cataloging-in-Publication Data:

Zonnenshain, Avigdor.
Managing and engineering complex technological systems / Avigdor Zonnenshain, Shuki Stauber.
pages cm
Includes index.
Summary: "This book is based on a qualitative study that includes dozens of in-depth interviews with experts in the systems engineering field"–Provided by publisher.
ISBN 978-1-119-06859-4 (hardback)
1. Systems engineering. 2. Industrial management. I. Stauber, Shuki. II. Title.
TA168.Z66 2015
620.0068–dc23

2015000703

Typeset in 10/12pt TimesLTStd by SPi Global, Chennai, India

Cover design by Zvi Fedelman.

Printed in the United States of America

10 9 8 7 6 5 4 3 2 1

1 2015

To my beloved wife, Michaela, for her love, friendship,
and continuous support

Avigdor

To Yaron and Sharona, my lovely offspring
Think before act, do not spare planning

Shuki

CONTENTS

WORDS FROM INCOSE PRESIDENT

Dear Systems Engineers,

I am proud to be a systems engineer.

Systems engineers have an important role in developing systems for the benefit of mankind and the world.

The study on Systems Engineering has Many Facets, which is documented in this book, is very important for the Systems Engineering community, as it presents for the first time different views and various aspects of systems engineering as expressed by 24 experts who have been interviewed for this research. These experts come from different countries, cultures, and backgrounds.

This study is a great asset for INCOSE as a training material and as a promotional document for the systems engineering community as it presents the diverse, rich, and colorful roles of the systems engineer.

I am proud to be part of this effort as one of the experts to be interviewed. It gave me the opportunity to communicate my views and vision on the important values of the systems engineer.

I congratulate the leaders of this initiative – Shuki and Avigdor, and I express my thanks to the Gordon Center for supporting this study.

John A. Thomas, ESEP
INCOSE President
(2012–2013)

WORDS FROM THE HEAD OF THE BERNARD M. GORDON CENTER FOR SYSTEMS ENGINEERING, TECHNION

In the early days of my term as dean of the Technion's Aerospace Engineering Faculty (1995–1998), I received a request from the industry to devise an academic program that taught systems engineering. Until that time, academic activity in systems engineering in Israel was fairly limited and spread out among different academic units of the Technion and other institutions. The fact that the industry officials had chosen the Dean of the Aeronautics and Space Faculty was no coincidence. Globally, systems engineering activity originated in the aerospace and defense industries. Being aware of the importance of this subject, and of the Technion's historical role as the promoter of engineering knowledge and achievement in Israel, I accepted the request without hesitation. The multidisciplinary and interdisciplinary nature of systems engineering made it clear that the program had to be an inter-faculty program – with more than one academic unit involved. Indeed, since its founding, the program has been managed by a committee of representatives from several different faculties. Moreover, I recognized that systems engineering was a field that, probably more than any other, had close ties to the industry. In light of this, the committee founded to determine the objectives and character of the studies included high-ranking industry representatives, in addition to its traditional academician members. The plan devised by the committee, even in its early versions, was for the program to grant its graduates a master's degree. Candidates had to have a bachelor's degree and a certain minimum of professional experience in their field of expertise, before they were even considered for the program.

After its launch, the program received a lot of interest, and demand was much higher than we had expected. Today, less than 15 years since, the program boasts over 1000 graduates and students. The systems engineers trained by it are spread out throughout the Israeli defense and civilian industries, offices, and organizations. They have tremendous influence on the integration of systems engineering procedures and systems thinking in Israel. One would be hard-pressed to find a major project in Israel that does not have graduates of the program among its teams, many of them occupying high ranking positions.

A considerable boost to systems engineering in the Technion and in the whole of Israel was brought on by the establishment of the Bernard M. Gordon Center for Systems Engineering. The center was founded thanks to the help and generous contribution of Dr. Gordon, a highly accomplished engineer from the United States, who recognized the importance of this cause. The center has allowed us to deepen the ties between the industry and the academy, and to begin research in systems engineering, some of it in close collaboration with the industry. Additionally, the center organizes conferences and seminars, invites foreign experts to hold lectures and workshops, and encourages the Israeli technological community to adopt modern systems engineering methodologies. Although, in recent years, awareness of the benefits of systems engineering has given it a firmer grasp in ever-growing parts of the industry, there are still engineering fields where it is virtually nonexistent. In light of this, the Gordon center believed it very important to perform the study that eventually led to the writing of this book. I believe that this book, based on interviews with local and foreign systems engineering experts, presents a broad view of the systems engineering field, the challenges it is facing, its contributions, its benefits, and the challenge it sets before those who wish to embed it into their industries and organizations.

Systems engineering has brought valuable insights on analyzing and optimizing the workings of complex, multidisciplinary and interdisciplinary systems, which can and should have an impact beyond the boundaries of the technological world. Health-care systems, social systems, and many other systems can benefit from adopting the methodologies systems engineering has developed. This is why it is so very important for those who are not involved with engineering or technology to be aware of and have access to systems engineering methodologies. With this end-goal in mind, the authors of this book have succeeded in presenting a picture that can be grasped by readers with no special technological background.

I would like to thank all the interviewees who have taken the time to participate in the interviews and helped put them in writing. I am grateful to Dr. Avigdor Zonnenshain and Mr. Shuki Stauber for their dedication to the writing of this book and for the great effort they have invested in it. I hope you find this book enjoyable, and I believe it will help spread the awareness of the many benefits of systems engineering. I hope this book will encourage its readers to use modern insights, knowledge, and tools to make better systems.

Prof. Aviv Rosen
Head of the Bernard M. Gordon Center for Systems Engineering,
Technion – IIT, Haifa

WORDS FROM THE PRESIDENT OF THE ISRAELI ASSOCIATION FOR SYSTEMS ENGINEERING-INCOSE_IL

Those who practice systems engineering have always intuitively known the right way to work on systems projects, but, until now, there was not enough quantitative data to support this perception. In recent years, studies have shown significant high correlations between investment in systems engineering (as part of a project's total cost) and project success, measured by meeting planned performance, budget, and schedule. High correlation was also found between the project's chief systems engineer's management and technological leadership abilities and the project's success.

It appears that the previous dilemma of whether to invest in systems engineering has now been resolved. Israel's defense industry and government organizations, the birthplaces of Israeli systems engineering, have long since understood this, and, in recent years, we have been witnessing the introduction of systems engineering approaches and processes into small- and medium-sized "civilian" enterprises as well.

Here, in the Israeli Society for Systems Engineering, we are convinced that in Israel, of all places, systems engineering should be treated as a strategic asset and as a discipline in which we hold a major relative advantage. For this reason, we decided that alongside strengthening systems engineering in defense organizations, and tightening the cooperation between industry, government, and academia, we should also act to assimilate the huge body of knowledge accumulated on the subject to all of Israeli industry. We intend to achieve this by research, education, and activities aimed at sharing knowledge of systems engineering.

This book fits in well with these objectives. It provides an exciting opportunity to experience a close encounter with a vast pool of knowledge and insights of 22 systems engineering experts.

The editors of this book are two leading experts: Dr. Avigdor Zonnenshain, a senior, experienced systems engineer, who recently retired from his position at Rafael and currently works at the Technion's Gordon Center for Systems Engineering, and Shuki Stauber, a reputable expert on management and author of numerous management books.

On behalf of the Israeli Society for Systems Engineering, I would like to thank them for their valuable contribution to the advancement of systems engineering in Israel.

Best regards,

Professor Moti Frank,
President of the Israeli Association for Systems Engineering – INCOSE_IL
(2013–2014)

WORDS FROM THE WRITERS

Systems Engineering is a dynamic discipline that changes and evolves constantly, adapting to the changes in its working environment. It is affected by factors like technological change, developments of interfaced disciplines, research findings, and lessons learned from experience in the industry, to name only a few.

In the 40 or so years of my career, I, too, have personally experienced various aspects of and perspectives on systems engineering and the applications of systemic approaches in the industry and academia. I have applied a holistic, systemic approach to a number of fields, including: systems safety and reliability, systems experiments, integrated logistic support, total quality management and engineering, management and development of systemic-technological projects, and corporate social responsibility.

These are the thoughts and experiences that led me to the idea of a qualitative study of the various aspects of systems engineering, carried out in the form of interviews with local and international systems engineering experts. To ensure the high level and professionalism of the interviews and writing, we employed the services of Shuki Stauber, a professional interviewer and writer, who specializes in management.

The interviews in this book will help the readers learn about the origins and evolution of the systems engineering discipline and gain a personal familiarity with systems engineering experts: their experience, opinions, and attitudes in this field. For this reason, we chose to call this study "The Many Faces of Systems Engineering." We sought answers to the questions: What are the different aspects of systems engineering? What different perspectives will the experts interviewed in this framework have to offer?

We approached over 20 experts, both locally and worldwide, representing a wide spectrum of occupations and experiences, both in the industry and in academia.

All experts responded to our request and agreed to participate in the study with great enthusiasm, presenting us with clear, detailed accounts of their experiences and opinions on this study's areas of interest, including answers to questions such as: How does systems engineering handle technological complexity and the ever-changing needs of the clients? How is systems engineering actually implemented in various projects and organizations? How does the systems engineer serve as both manager and leader? All the experts we approached applauded the initiative, were glad of the opportunity to present their approaches, and showed great appreciation for the possibility that their opinions shall be banded together and included in a book that would summarize the findings of the study.

It should be noted that my personal acquaintance with dozens of systems engineering experts in Israel and worldwide has helped us recruit this study's participants. In fact, we have yet to hear a single refusal …

This study has also helped me achieve some historical closures, one of them being my meeting with President of Lockheed Martin, Norman Augustine, whose acquaintance I had had the pleasure of making back at Martin Marietta, when he, as its president, had been navigating the collaboration with Rafael, and I had been a member of the team that facilitated its establishment. Back then, he had been the object of my admiration, and now, I have had the privilege of meeting him and interviewing him for this study.

The choice to recruit Shuki Stauber, a professional writer and interviewer specializing in management fields, to help carry out this study, was both conscious and educated. We wanted the interviews to be very professional and the writing of their summaries to be accessible to all. This meant that readers from outside the systems engineering community should be able to read and understand the findings and insights contained herein and even apply them to their respective occupational areas. It is my opinion that the "gamble" of recruiting Shuki has paid off beyond our wildest expectations: Shuki listened to and learned the various aspects of systems engineering and has managed to put the findings into words that anyone would be able to understand. I can say without hyperbole that Shuki Stauber is the best systems engineer among all the authors of management books in Israel.

I would like to thank the head of The Gordon Center for Systems Engineering at The Israel Institute of Technology (Technion), Professor Aviv Rosen, for bravely taking up the gauntlet I had tossed his way, in the form of this extensive study. Professor Aviv Rosen handled the financing of the study, directed its movements, and actively participated in most of the interviews. His professional contribution to this study is nothing short of priceless.

I hope that you find this book interesting and enjoyable.

Dr. Avigdor Zonnenshain

When Dr. Avigdor Zonnenshain first approached me and suggested that we coauthor a book on systems engineering, I was not even familiar with the term. Neither "engineering" nor "systems" sounded like exciting words to me. However, after he had explained what it was all about, I found the subject interesting, special, and innovative.

I had hardly had any previous contact with technological fields. I acquired my basic education in the Faculty of Social Sciences, specializing in "soft" areas, centered on human skills. As a management expert, my professional writing focused on connections between people, human abilities, and the organizational world.

I suddenly discovered that engineers, pure technologists, are beginning to understand that technology does not exist in a vacuum; that it is meant to serve people. In the words of Professor Aviv Rosen, an aeronautical engineer and currently the head of The Gordon Center for Systems Engineering at The Israel Institute of Technology (Technion), which provided the framework for the study this book is based on: "I came from the world of exact science, and I have come to see that in many cases, the 'soft' sciences are no less important than the technology itself. There is always a client at the end of the road, and the user's psychology must be taken into account."

Commonly attributed to Plato, the proverb "necessity is the mother of invention" is very relevant to systems engineering, a field that emerged from the need to deal with the increasing complexity of technological systems. I have found that methodologies developed in order to cope with technological-systemic constraints can also serve managers, whose very nature is to handle managerial-systemic constraints. One way or the other, the gap between these two worlds, the technological and the human, is closing – everything connects to everything else, as we head toward the formation of supersystems, combining technology and people together.

Indeed, while working on the book, I have found excellent tools that can serve not only the managerial needs of engineering but also all the worlds of management in general. This book is, therefore, a book on management for all intents and purposes. It is intended for managers as well as systems engineers, in equal measure. It has been written in this form, so that managers with no technological background can derive valuable knowledge from it too.

Finally, I wish to express my sincere thanks to my two colleagues, Professor Aviv Rosen and Dr. Avigdor Zonnenshain, who have given me the opportunity to be a part of an exciting process of learning and creation. Besides being first-class professionals, their extensive intellectual abilities are combined with kindness and good-naturedness – a wonderful mix that has made working with them an extraordinary experience.

Shuki Stauber

PREFACE

SYSTEMS ENGINEERING – A DISCIPLINE IN THE MAKING

A discipline in the making, systems engineering connects classical engineering with organizational and managerial systems. It is, therefore, not surprising that one of the main skill sets required of a systems engineer (and its importance increases as the engineer's career progresses) is that of leadership skills. This fact stands in contrast with systems engineering being, at its core, an engineering discipline, practiced by engineers.

In the past, there had been a clearer distinction between professional engineers, skilled in their fields (for instance, electronic engineers, mechanical engineers, or computer engineers) and integration and management people, who brought together the technological systems developed by the engineers. But the ever-accelerating technological developments and globalization created a situation, where this separation delayed processes and compromised the development abilities and, consecutively, competitiveness of those organizations that unconsciously kept sanctifying the distinction between these two overarching areas.

For example, in the past, an engineer could demand to go back and perform countless tests in order to achieve technological perfection. But today, he must take into account such considerations as resource availability and scheduling. *He can no longer act based on "pure" engineering considerations. Being forced to face other, systemic constraints, as well as the "traditional" challenges of his occupation* means he must now think like a systems engineer.

Not all engineers have to undergo this transformation. Many want to continue to focus exclusively on their professional area of expertise. Some, however, wish to use the engineering analysis tools they had acquired (both in their academic studies and during their work) not only to develop a sophisticated electronic circuit, an advanced machine, or a complex piece of software, but also to "engineer" an entire system that enfolds not only technology, but other components as well, including economic constraints, human factors, and commercial and marketing considerations – these are the systems engineers.

Systems engineers adopt managerial thought patterns, because technology can no longer be kept separate from the wider context it exists in. They gravitate towards systems that rely heavily on technology, where they are best able to utilize their relative advantage: they enhance their managerial thought using the engineering analysis tools they had acquired. This is why so many systems engineers can be found in organizations that develop aircraft or advanced weapon systems, while very few (if any) work for financial institutions or retail networks.

This book attempts to open a window into the world of systems engineers, allowing the readers to learn from the accumulated experience of the people interviewed for the study and perhaps help them adopt new thought patterns and methods of conduct, suitable to each reader's area of activity. After all, system engineering is, by its very nature, a discipline that traverses the gaps between other disciplines, and so, its methodologies can serve a wide variety of experts from the worlds of management and engineering – from the production manager, wishing to improve his production line; to the medical doctor, developing a new type of syringe; the architect, designing a complex of buildings; and, finally, to the head of a government office, formulating a multiannual plan.

On the structure of this book:

Managing and Engineering Complex Technological Systems is based on a qualitative study that included, at its core, dozens of in-depth interviews with prominent experts in the field. We have conversed with lead systems engineers, high-ranking executives, academic experts, and experienced consultants. Roughly one half of the study's participants are Israelis, with the other half hailing from a multitude of other countries. We felt it was important to present a wide spectrum of perspectives on this field, as it pertains to the different features of various industries and organizations.

The first part of the book is a general overview of the systems engineering field. It includes its origins and the history of its emergence, its main characteristics, and the directions of its evolution. This overview is based on a series of interviews with the experts and the insights derived from the discussions with them.

Systems engineering grew out of needs that had arisen in complex technological projects, and it maintains a heavy presence in these frameworks to this day. We demonstrate this in the second part of the book, where we discuss two significant Israeli defense system projects. The first is the IAI Lavi project, launched over 30 years ago and never completed; the second is The Iron Dome project, completed successfully and highly acclaimed by the media. The comparison between the conduct of these two projects is very informative, especially when it comes to systems engineering, seeing as, among other things, it illustrates the substantial changes the discipline had undergone during the years that passed between the two projects.

The third, and final, part of the book contains the detailed interviews themselves. It includes rich, detailed information that relies on the knowledge and years of experience accumulated by the people we met with. These chapters are divided into five sections:

- Systems engineering as the answer to the challenges of a complex technological world – the aerospace industries;
- The development of systems engineering in the commercial and industrial worlds, and in complex civil systems;
- The impact of the accelerated development of the computing world on systems engineering processes;
- Systems engineering and the academic world;
- Systems engineering in the world of training and consulting.

Finally, we would like to thank all the experts who contributed to this book, shared their wisdom and knowledge, and gave us their time. Their names and professional backgrounds are listed in the acknowledgements section, on the next page.

AVIGDOR ZONNENSHAIN
SHUKI STAUBER

LIST OF INTERVIEWEES (ALPHABETICAL ORDER)

AA
Chief Systems Engineer at Rafael Advanced Defense Systems Ltd.

Yossi Ackerman
President and CEO of Elbit Systems Ltd. (1996–2013).

Norman Augustine
President, CEO, and Chairman of the Board of Directors of Lockheed Martin (1987–1997); has since served as the chairman of several presidential and national committees, including US Antarctic Program Blue Ribbon Panel for the assessment of the United States' activities in the South Pole.

Henry Broodney
Head of the Systems Engineering Technologies Group at the IBM Research Lab in Haifa.

Boaz Dovrin
Project manager at Luminis; formerly systems engineer and technical manager at Elbit Systems and Philips Medical Systems.

Sanford (Sandy) Friedenthal
International expert, lecturer, and consultant on model-based systems engineering (MBSE); formerly a Lockheed Martin Fellow.

Dr. Gilead Fortuna
Senior Research Fellow, head of the project "Israel 2028 – Vision and Strategy for Israel" at The Samuel Neaman Institute (since 2009); formerly Senior Vice President at Raphael Advanced Defense Systems Ltd. and Teva Pharmaceutical Industries Ltd.

Alon Gazit
Director of R&D; **Benjamin (Benjie) Rom**, Head of Product Development; **Erez Heisdorf**, Head of Eilat Program – HP Indigo

Dr. Ovadia Harari
Head of the Lavi Project (1980–1988), Vice President of Israel Aircraft Industries Ltd. (until 2006), winner of the Israel Defense Prize for the years 1973 and 1969, winner of The Israel Prize for Technology and Engineering for the year 1987.

Dr. Cecilia Haskins
Associate Professor at the Norwegian University of Science and Technology (NTNU), ESEP, and member of the INCOSE Board of Directors.

Dr. Eric Honour
International expert, consultant, researcher, and teacher of systems engineering.

Prof. Joseph Kasser
International expert, lecturer, and researcher on systems engineering; currently, a visiting professor at the National University of Singapore (NUS).

Harold (Bud) Lawson
International expert, researcher, lecturer, and consultant on Systems Thinking, Systems Engineering, and Software Engineering; Professor Emeritus of telecommunications and computer systems at the Linkoping University, Sweden.

Niels Malotaux
International consultant, specializing in helping projects and organizations.

Dr. Jacob (Kobi) Reiner
Chief Systems Engineer at Rafael Advanced Defense Systems Ltd.

Prof. Aviv Rosen
Faculty member at The Faculty of Aerospace Engineering at Israel Institute of Technology (Technion), initiator and current head of the Technion's Systems Engineering ME program, also, currently, the head of The Technion's Gordon Center for Systems Engineering.

Sharon Shoshany Tavory
Systems engineering consultant, researcher at Gordon Center for Systems Engineering; formerly Head of Administration and Chief Systems Engineer at Rafael Advanced Defense Systems Ltd.

Hillary Sillitto
Systems Engineering Director and Thales Fellow, Thales, UK.

John Thomas
President of the International Council on Systems Engineering – INCOSE – (2012–2014), Senior Vice President and Chief Systems Engineer at Booz Allen Hamilton consulting firm.

Miriam (Mimi) Timnat
Senior Systems Engineer, Head of Process Improvement in Systems Engineering and Technical Management at Elbit Systems.

Prof. Olivier de Weck
Faculty member of the Engineering Systems Division at MIT, editor-in-chief of the journal *Systems Engineering*; formerly, an officer in the Swiss Air Force.

Dr. Amir Ziv-Av
Chief Scientist at The Ministry of Transportation (2011–2013), founder and owner of an engineering-technological company.

PART I

SYSTEMS ENGINEERING – A GENERAL OVERVIEW

1.1

THE ORIGINS, HISTORY, AND UNIQUENESS OF SYSTEMS ENGINEERING

For many decades, each of the industries that relied heavily on engineering, such as electronics, mechanics, and chemistry, had its own unique discipline. The engineers of each discipline evolved and gained experience in their respective specializations. *But, in the early 1970s, the need arose to integrate the various engineering fields and even bridge the gap between engineering, as a whole, and nonengineering systems.*

This phenomenon has its source in two opposing trends: on the one hand, engineering disciplines were becoming more and more specialized; and on the other hand, the need for multidisciplinary skills was on the rise.

Clarification: technological developments led to an increase in specialization and created a need for more and more specialists in subdisciplines of engineering. Today, for instance, an electronics engineer would not be considered an expert in electronics, but rather in one of its more specific subdisciplines, such as communications or control systems. Therefore, in order to create an electronic system, one now needs to integrate these subdisciplines. On the other side of the spectrum, the technological capability of manufacturing complex products for the end user's benefit raised the importance of integration between the overarching engineering disciplines, such as mechanics, electronics, or materials engineering, as well. At the same time, the systemic complexity of developed systems began to increase too.

This need gained much momentum from the development of the software and software engineering fields. Software allows for the creation of complex systems and is,

Managing and Engineering Complex Technological Systems, First Edition.
Avigdor Zonnenshain and Shuki Stauber.

in many cases, the central factor that facilitates the combination of subsystems from various disciplines.

In the words of Eric Honour, who believes that the evolution of systems engineering in the 1980s received a substantial booster shot from the breakthrough in the software field that took place during those years: "In the late 1980s, the industry was faced with a major problem: many software failures were discovered, because software personnel had not received the information they needed, at the quality they required. They began looking at ways of receiving better requirement specifications, and that put systems engineering back into people's minds."

We can assume that one of the main reasons for the emergence of systems engineering is the development of technological abilities that, with the help of software, allowed for the creation of technological systems of ever-increasing complexity. This phenomenon created the need for a technological position holder, charged with the task of integrating the subsystems that form the complex, overarching system.

Technological systems are an integral part of the modern world. They provide us with a variety of services, and play a part in larger and larger systems, some of which contain nontechnological components as well. *This process leads to the creation of supersystems, which can no longer be effectively controlled*, because they are entwined with human systems, among other reasons.

These needs and constraints greatly increased the need for the formulation of an orderly, methodological, systematic approach to the management of complex engineering systems. *The importance of early planning rose greatly*, as did the need for skills that focus on areas other than "pure" engineering. Hence, the technological industry began to understand the increasing need for engineers who followed a predetermined, orderly, controlled, and supervised methodology: *a methodology that would allow them to design holistically and facilitate educated integration processes, while minimizing the ever-present, ever-increasing risk of failure*. Failures were brought on by loss of control over large, complex systems, exposed to a wide variety of constraints, some of which were organic and process-related, rather than pertaining to engineering.

On top of these, came the *financial component*: as aforesaid, the ever-increasing engineering capabilities opened the possibility of developing more and more sophisticated products. As engineering projects grew in scope and complexity, so did their financial costs. Uncertainty levels rose. Complicated projects failed to meet deadlines, went over-budget and even got canceled (a prominent example of this is Israel's Lavi project, which we discuss later in this book).

These trends brought about significant changes in the way large engineering projects (wherein, as aforesaid, the use of systems engineering is especially important) were approached. For instance, in the past, governments used to allocate nearly unlimited resources to the development of complex defense products, such as fighter planes. Today, the costs of these products are so high, that no government can reasonably afford to invest in this area without strict budget limitations and control. Before the late 1970s, most defense projects were managed using cost-plus pricing strategy, while today, most projects operate on a rigid, given budget and are hardly ever allowed to deviate from it. This led to the emergence of yet another fundamental constraint that forced engineers to consider nonengineering factors, especially when planning and developing complex engineering products. It should, however, be noted that the adoption of work methods that rely on a predetermined budget was no magic cure-all, and today's large, complex projects still fail to meet deadlines and stay within their budgets.

Our choice to bring the development of a fighter aircraft as an example of a complex engineering project was not incidental. Some say that aeronautics may be the most complex technological field of all (due to the need to control a large, complex, man-carrying, airborne vehicle). This is what caused systems engineering to evolve in this technological area, first. Moreover, many aeronautics experts contend that they have always been employing the principles of this approach; only back then, they did not refer to it as "systems engineering." As the need for systems engineering became more urgent and began to seep into other industries, it slowly gained recognition as an independent discipline – one worthy of its own, separate training program and career development paths.

1.1.1 ON THE ESSENCE OF SYSTEMS ENGINEERING

Being a discipline in the making, there is, as yet, no consensus on the character and operational frameworks of systems engineering. The interviews in this book show a myriad of perspectives regarding not only the nature of this profession, but also the question of whether it is indeed just that – a profession.

Yossi Ackerman says that a systems engineer is a vague term that defies definition, and he is glad for it. This is because, according to him, this vagueness creates a flexibility that allows the job to be adjusted to suit the circumstances. Ackerman sees systems engineers as "managers, who are engineers by profession, but are able to see the whole technical, technological picture." The more senior the engineer, the more management-oriented his job becomes.

Many experts find that systems engineering is more than a job, it is also a collection of thought and work patterns. Thus, for instance, Mimi Timnat finds that "to a great extent, *systems engineering is more than just a job*. It is an approach to handling and solving problems, and not only work-related ones. It encourages one to look at a problem from different angles, to ask questions and try to gain a better understanding of the problem, before making decisions and formulating solutions." Dr. Cecilia Haskins goes even farther, believing that *a systems engineer does not have to be an engineer at all*, and the word "engineering" may have been wrongfully applied to this term. According to her, the important component is the "systematism" – the ability to see the whole picture and perform the necessary actions methodically. In her perspective, systems engineering is a combination of discipline, worldview, and profession that suggests ways of solving problems.

Niels Malotaux expands on this issue: "*All engineers must be capable of systems thinking.* There is no point in completing part of a system, if it doesn't work together with the other parts. I do not view myself as a systems engineer, although I meet the definition of one. The principles of systems engineers lie at the heart of all engineering disciplines, and all engineers should be able to find an optimal compromise between opposing requirements. 'Systems engineering' is a label applied to all the things engineers have to do, in order to create a good system."

1.1.2 THE DIFFERENT TYPES OF SYSTEMS ENGINEERING

Even the most enthusiastic supporters of systems engineers do not think there is such a thing as a "pure" systems engineer. They believe *systems engineers are engineers who have gained knowledge and experience in one of the classic engineering disciplines* and, being people who possess certain skills and character traits, were able to grow into systems engineers. This professional growth pattern is commonly illustrated using a T model. This model, named for its shape, which resembles the letter "T," places basic training in a concrete engineering field on the vertical axis, and the lateral, multidisciplinary view of the systems engineer, on the horizontal axis.

This perspective suggests that academic programs that train systems engineers have to be for a master's, rather than a bachelor's, degree, and in order to be accepted, applicants would have to possess an engineering degree in one of the fundamental areas of engineering, as well as some hands-on experience working as an engineer is recommended.

Tendencies toward systems engineering are more common among engineers, whose areas of technological expertise frequently involve the tasks of *examining alternatives and facilitating integration.* Aeronautical engineers, whom we have previously mentioned in this context, fit this pattern well. Electronics engineers also have similar traits.

Systems engineering also has some managerial traits, which we will discuss in detail later in this book (see Chapter E), and just as there are different types of managers, there are also different types of systems engineers. *Two super types* prominently figured in the interviews we had performed, their identity derived from the fundamental essence of systems engineering, *as a method that allows its practitioners to analyze a system to its smallest details, and then design it to suit the client's needs.*

This process is commonly illustrated using a "V" shaped graph, where the horizontal axis is time, and the vertical axis is the level of detail. The first step of developing a system is to define the needs of the client. This is represented by the highest point on the V's left branch. Next, begins the process of generally designing the system as a whole, and delving into the details of its subsystems, which should provide concrete answers to the previously defined needs. This "descent" to the lowest point of the V reaches the characterization of the subsystems' most basic components, an activity that focuses on system *analysis*. From here on out, we begin to "ascend" the right branch of the V. This includes the product development and testing processes, all the way to completion. This activity is known as system *synthesis.*

The analysis and synthesis are based on different action patterns that require the use of different skills. Analytic systems engineers are tasked with development, design, and architecture, while synthetic systems engineers focus on implementation and integration.

Prof. Aviv Rosen explains: "Analysis is associated with the world of research, and its products are usually models for understanding various phenomena. Innovations often begin with analysis. Conversely, synthesis is the ability to bring components together and produce an engineering product. This is usually done by the industry. Synthesis is considered to be of a more routine nature, and was therefore perceived as inferior to analysis by the academy for many years. Research was thought of as a more lucrative practice, as it offered the possibility of discovering new things and publishing one's findings in scientific magazines. But times have changed, and the importance of synthesis has slowly increased in many engineering fields. Consequently, the rate of appearance of major technological innovations in these fields has diminished."

Henry Broodney also distinguishes between two types of systems engineering: he defines the first type as systems engineering that deals with planning and the process of designing the system itself. It focuses on what the project's lead systems engineer does, and Broodney refers to it as *Technical Systems Engineering*. The second type of systems engineering includes processes and work methods for managing engineering projects that combine scope, scheduling, and finances. The focus is on the actions of the project manager – on managing the system. This, he calls *Management-Oriented Systems Engineering*.

These two broad classifications are likely to include various systems engineering "subdisciplines," certainly in more complex projects. For instance, technical systems engineering includes systems engineers tasked with developing a certain component that pertains to a field of engineering they are well-versed in (e.g., a systems engineer trained in electronics engineering, in charge of developing an electro-optical component), and systems engineers tasked with integrating components from various disciplines, in the process of creating a comprehensive technological system that meets the client's needs. For example, The Iron Dome project employed systems engineers from all areas, including software, electronics, and mechanics, who worked alongside two systems engineers who coordinated all technological activity, within the framework of what AA referred to as "Lateral Systems Engineering." These two were interdisciplinary systems engineers, while the other systems engineers in the project worked within the boundaries of their specializations.

1.2

A MULTIDISCIPLINARY, SYSTEMIC VIEW

In the previous chapter, we discussed the trends that led to the increase in the need for systems engineering, namely: *the ever-growing complexity of technological systems*, alongside the ever-increasing demand for *appropriate solutions for the needs of the clients* – the buyers and users of the systems (the two are not always the same). This combination compels engineering teams charged with the development of technological products to account for nontechnological constraints, related to finances and management. Therefore, the systems engineer who manages these teams should be willing to engage in areas beyond his formal engineering training, in the desire to meet the clients' needs, by exposing himself to broader technological fields, while handling managerial and organizational issues.

On the increasing complexity of systems:

The continuous accumulation of knowledge allows for the creation of advanced systems that are, naturally, also very complex. It is a well-established fact that the more complex the system, the higher the risk of it being prone to faults and difficult to operate. It follows that one of the main challenges is creating a product that is as technologically advanced and, at the same time, as simple as it can be. This increases the need for *strong simplification capabilities* alongside *efficient examination of alternatives*.

The Iron Dome developers attested to this: "We could have gotten a more complex 'servo' in one round of development. To reach a 'servo' that simple required more thought and more talent. Finding a complicated solution is fairly easy. To find a simple solution, you need to start thinking."

Managing and Engineering Complex Technological Systems, First Edition.
Avigdor Zonnenshain and Shuki Stauber.

On the increasing importance attributed to clients' needs:

The last few decades are characterized by the increasing power of clients everywhere: from fashion retail to education and healthcare services. This trend did not skip the technological world, where, often, the client's representatives are no less competent than the developers they meet with. Clients' control systems for complex technological products have grown tremendously, and their involvement in all stages of product development has increased. The need to account for the clients' needs and demands has become paramount. Among other factors, this trend is the result of the changes in budgeting methods, as clients are now much less lenient when it comes to deviating from predetermined financial frameworks, and so grew the demand for engineers who knew *how to handle themselves with the clients' representatives*, who were able to negotiate with the clients and speak their language.

In many cases, clients see themselves as the ones making the demands, and the developers as the ones tasked with meeting them. This pattern of conduct tends to have a negative impact on the work rate and even on the quality of the final product. Today, the world is beginning to realize that this fine weave of relations has to be handled wisely and with care, which brings us back to The Iron Dome Project, whose developers stated that in their case, "The client almost merged with the project. This does not go without saying, and there are those who even criticize it, saying that perhaps it is best for the client to keep his distance, so that he may represent the other side, and maintain his ability to provide objective feedback. In The Iron Dome Project, however, it worked very well, because of the client's representatives' ability to successfully maintain their independent thought."

1.2.1 THE BOUNDARIES OF A SYSTEM

One of the major dilemmas encountered by all who practice systems engineering, including the systems engineers themselves, is the question of defining a system's boundaries. Thus, for instance, one might decide that the boundaries of a system are the client's technological requirements of the project. But one might also decide to include the technological system's impact on the environment. Naturally, such decisions can radically change the design and character of the system.

Expanding the boundaries of a system also reveals context that cannot always be seen from within the system's original boundaries.

Prof. Olivier De Weck: "We design a system with certain boundaries and see no correlation between A and B, but if we expand the boundaries, we can suddenly see a correlation (or synchronization) outside the original system. Seeing this correlation has a profound effect on how the inside of the system is designed. For this, the concept must be modeled in greater detail."

Prof. Joe Kasser stresses this as well: "Defining the boundaries of the system is critical. For one person, the system is the car; for another, it is the car and its passengers; for a third person, all the cars on the road are the system. The engine is also a system. This is where systematic thinking is needed."

1.2.2 SYSTEMS OF SYSTEMS

Olivier De Weck says that "traditional systems engineering had always been inward focused. It made sure all the system's components (the components, the processes, the subsystems) worked together to produce the system's end products and satisfy the requirements set forth by the client. Systems engineering never gave much importance to what lay outside the system. Today, many systems are starting to turn into '**systems of systems**'. They become very large and more and more complex. We start connecting systems that were not designed to work together."

Prof. Hillary Sillitto expands on the importance of *systems of systems*: "We connect systems together, creating mega-systems, because they allow us to do things we could not do otherwise, to solve problems that cannot be solved otherwise. They allow us to do things better, or to develop new business models and create new opportunities, like the Internet has.

There is a 'super problem' that stems from the formation of such enormous systems: the large number of risks these systems entail. When designing complex systems, the thinkers and planners see the opportunities and chances, but are not always able to assess the risks and try to minimize them early, at the planning and design phase."

Olivier De Weck gives an example from the field of transportation: "Drivers texting behind the wheel is currently the most common cause of traffic accidents in the United States. This has to do with systems engineering, because if you analyze the problem, the traditional transportation system is now joined with communication systems in unexpected ways and means of communication, with human behavior and motivation at the center of it all."

Sillitto says: "The more complex systems become, the more the connections between them multiply, the higher the chances that something will go wrong, be it on purpose or due to plain stupidity. Thus, the importance of the need to balance opportunities and risks cannot be stressed enough."

The multileveled nature and complexity of systems raise the importance of *risk management* in all systems, whether they pertain to engineering or not. Dr. Gillie Fortuna gives an example from a system that constitutes the organizational structure of a major corporation, spread out around the globe – Teva Pharmaceutical Industries. Efficiency considerations have led Teva to place its marketing and production arrays under different managements. This created "a lot of interdependency between subsystems. The ability to manage client commitments without controlling the resources entails a lot of risk management that only a fine-tuned cooperation between all the involved factors can achieve."

Dr. Amir Ziv-Av also raises the importance of *optimization*, which he defines as: "Viewing the system as a whole, in its ensemble of economic, operational and technological components." According to him, "a 'product' is an answer to a collection of differently weighed objectives, and at the heart of its development process stands the task of maximizing the target function. In the end, to win the competition over the heart of the client, one must have a relative advantage, which is attained by doing more with fewer resources."

1.2.3 MANAGING THE HUMAN FACTOR

The increasing complexity of technological systems has, in turn, impacted the *complexity of organizational systems*, where systems engineering operates. Developing a complex technological system requires the skills of many people, hailing from many different fields. Therefore, *the systems engineer's integration work does not end with the technological context; today, more than ever, it is required in the human context as well.* More and more experts believe that the importance of the ability to lead multidisciplinary teams outweighs even that of the systems engineer's technological competence.

This need has grown even further due to globalization, which played a key role in the increase in the power and importance of multinational corporations. These companies wish to utilize their advantages of size and global deployment by running multinational, multidisciplinary work teams, spread out around the globe. Of course, this approach makes the task of directing them even more difficult.

1.2.4 TRAITS DERIVED FROM AN INTERDISCIPLINARY, SYSTEMIC VIEW

The aforementioned shows just how crucial it is for a systems engineer to possess *an all-inclusive, systemic view.*

Boaz Dovrin is of the opinion that a key condition for a systems engineer's success is his ability to visualize the end of the project from day one. Prof. Ovadia Harari explains the necessity of this trait from the opposite direction as well: "a systems engineer must see the whole picture and use common sense to filter out the less important details, otherwise, the endless dive into the small details will disrupt his work processes."

An all-inclusive, systemic view cannot exist without a second important ability: *multidisciplinarity*. In the words of Prof. Aviv Rosen: "systems engineers are people who are willing to delve into areas outside their natural habitat."

Ovadia Harari illustrates this statement: "a systems engineer should take the budget issue into account. He must understand that money is a vital parameter. If a systems engineer is focused solely on technology, he will not have the correct balance, required of a good systems engineer." He also adds that "a systems engineer should be able to *talk to a variety of experts in a clear and simple language.*"

The ability to simplify is important, not only for dialogue with different experts, but also to provide answers to technological needs. John Thomas believes that the ability to simplify things, "to see beyond technology and understand the problems," allows a systems engineer to fulfill one of his most important missions – *problem solving*. John Thomas also uses the word "audacity" to describe another trait required of systems engineers. A systems engineer must be audacious, to be able to strive for the accomplishment of his tasks, while breaking through the obstacles in his path and dealing with limitations and difficulties.

Other experts also see *systems thinking as a way to solve problems*. Sandy Friedenthal, for instance, says he applies systems thinking to everything he does: "it's a way of thinking, an approach to problem solving. The focus of this approach is to understand different stakeholder perspectives and concerns, and define a problem first before jumping to a solution. Then establish value from the perspective of the stakeholders, determine alternative approaches to address the problem, evaluate the alternative solutions, and validate the solution addresses the need. System thinking provides a way to think about how the pieces of the solution fit together to address the problem."

Another important trait all systems engineers should possess *is the ability to choose between alternatives*. This is important, because "today's infinite (technological) possibilities" create a wide range of options on the one hand and a large number of constraints on the other hand – a situation that forces one to choose wisely.

Norman Augustine comments on this: "compared to engineers, who solve engineering problems, *systems engineers face problems that have intrinsic conflicts and numerous components*. They do this by analyzing and by making trade-offs: this is *a process of balancing out different considerations that also affects the determination of the appropriate 'dose' of each component in the system*. This is one of systems engineering's most important areas of activity."

In the same context, Dr. Kobi Reiner adds: "a systems engineer has to be able to cut. Engineers tend to complicate things, and it is his job to stop them, because today, the possibilities are endless. One of the key traits a successful systems engineer must possess is *the ability to simplify, when the atmosphere is one of complexity and complication*. A good systems engineer prevents complication from emerging."

These traits allow the systems engineer to shape the craft of *coordination and integration* between the subsystems that make up the system he is in charge of.

For a systems engineer to have these abilities, required for successfully doing his job, he must first possess a row of fundamental traits. Ovadia Harari presents three such traits:

The first is a mix of *openness, curiosity, and refusal to accept things as they are*: "systems engineers are dynamic individuals, involved in many different areas … they are open-minded; they ask questions and have a dialogue with you (as a manager). If you want something from them, they don't just go and do it, they ask you what you need it for. They sit with you and examine whether a different solution is also acceptable … "

The second trait is *learning ability*: "a systems engineer needs to be able to learn, to expand his mind, even at the age of 45. For example, the client raises a demand for an electric flight control system. You cannot tell him: 'no. I will revert to a mechanic flight control system, because that is what we know.' You open up the books, you take some courses, and you begin to understand."

The third trait is methodicalness: "a systems engineer is a methodological person. He has a method and the tools to use it. He must adopt process-oriented thinking."

Dr. Gillie Fortuna adds: "A good systems engineer has to see both ways. He must understand the subsystems and their limitations. This is a critical component of systems engineering and of a systemic view. This is why it is not enough to see things from the top down, you have to be able to see bottom up as well … a manager must maneuver between the need to focus and make operative decisions, and the ability to adopt a broad perspective of the consequences of the realization of the vision, and the path that leads to it."

The larger and more complex the system is, the more important the management mechanisms that control it. This effect reaches so far down the hierarchy that today, more than ever before, systems engineers are required to possess management skills as well.

Prof. Joe Kasser finds that the two areas, management and systems engineering, are intertwined: "There is an overlap between systems engineering and management, because systems engineering designs the processes that managers later supervise. There are a lot of professionals who use systems engineering tools. Systems engineering is a problem-solving mechanism that includes many managerial elements, because it involves human components and processes. It is a mechanism adopted mostly by engineers, because they found it useful for solving systemic technological problems."

The next chapter is dedicated to this issue precisely. In it, we will discuss one of the most vital overarching traits a systems engineer must possess – leadership skills and management abilities.

1.3

THE SYSTEMS ENGINEER AS MANAGER AND LEADER

We began the foreword to this book with the statement that "as a discipline in the making, systems engineering connects between classical engineering and organizational and management-oriented systems." Now, we seek to examine the combination of these two content worlds: one technical, physical, and accurate; the other behavioral and amorphous.

Prof. Joe Kasser believes that *systems engineering is not a profession, but a discipline*, a collection of work patterns. In his eyes, "systems engineering is the management tool of the 21st century; a different management method that includes tools and techniques suited for each case." According to this perception, nonengineers can adopt these work patterns too.

Not all the experts we have spoken with support this position.

Ovadia Harari contends that "*a systems engineer is, first and foremost, a technical man, who has to deal with lateral, technical, management issues. He must combine engineering skills with management abilities*. He cannot succeed without the combination of these two components."

Norman Augustine may agree with the statement that systems engineering is a management tool, but he stresses that "*systems engineering is more engineering than management*, it is a type of engineering that can handle 'non-physical' matters as well." He adds that, because systems engineering often includes more than mere technical skills, many engineers are frustrated by it, having no desire to handle "human" issues. Norman Augustine attributes great importance to the systems

Managing and Engineering Complex Technological Systems, First Edition.
Avigdor Zonnenshain and Shuki Stauber.

engineer's leadership and the interpersonal skills required of him: he must evoke trust, be a man of vision, be brave in his decisions and deeds, be capable and professional, energetic, and motivated; these are the traits that make him a worthy role model for his people.

In this context, Kobi Reiner distinguishes between two types of systems engineers: he, too, agrees with the statement that systems engineering contains management components. According to him, this is especially prominent in the systems engineering of projects, which, by their very nature, seek to arrive at a specific goal. Compared to that, systems engineers who operate in professional engineering units (that are often engaged in the development and support of projects – the authors) invest all their energy in engineering.

It appears that one issue remains undisputed: *a systems engineer needs leadership skills*. Some even see this as a key condition to his success.

The International Conference of the Israeli Systems Engineering Association, INCOSE_IL, which took place in Israel in 2011, housed a discussion of this very issue, with high-ranking executives in the industry. In it, it was said that "the traits of a leader overlap with those of a systems engineer. Thus, systems engineering is engineering leadership … *a systems engineer is an integrator of people, who hail from different disciplines, to advance a technological project*. He must, therefore, be the leader of the project, and not necessarily and engineering leader … (In this case) it is not the professional aspect that matters most, but rather the ability to lead people."

Prof. Aviv Rosen supports this approach: "To become a systems engineer, one must possess inborn traits. Leadership, for instance, is imperative."

Ovadia Harari adds: "Leadership and teamwork are more important for a systems engineer than professional leadership. He needs to lead people and handle crises. In these situations, soft qualities are vital; otherwise, people simply will not follow him."

Within the set of leadership qualities, the experts give substantial weight to interpersonal skills.

John Thomas: "Systems engineering is not management, certainly not. I (as a senior systems engineer) am not a manager, I am a leader. I allocate managerial responsibilities to others. A leader creates situations, where people want to follow him. He must be authentic and conduct himself with transparency. It is not enough to show and explain, I set the standards for my behavior, and by doing so, show others how I expect them to behave. This is how I instill the rules of systems engineering in them."

Ovadia Harari: "*Systems engineering is people-oriented in essence* ('people-oriented' is a term taken from a management model that places managers on a scale: at one end, stands a 'people-oriented' manager, while at the other stands a 'task-oriented' manager – the authors). If you cannot share and seek advice, you have failed as a systems engineer. You must convince your employee that your way is the right way. You have to compromise, otherwise people become small-minded. When you do not let people express themselves, they close up. Even if they have good ideas, they do not express them. They say to themselves: 'this manager has already made his decision; he doesn't want to be confused with facts'. *A systems engineer has to be a people person. A sociopath cannot be a systems engineer.*"

Kobi Reiner: "One of my main goals was to get to the end of a project with minimal stomach-aches on the developers' side. It is important to hear what is on their minds. It gives them a good feeling. I never gave the developers instructions without going through the leader, but I approached them to hear the goings-on."

AA from The Iron Dome project gave a fine example of the importance of these qualities, when he told us that the project's lead systems engineer had "Extraordinary Skills." When asked what those skills were, it was not the technical skills he emphasized. Rather, he said: "The ability to completely separate his professional agenda from his ego; although he has exceptional professional capabilities, he never becomes entrenched in prejudice. As a systems engineer, this approach allows him to have a dialogue with a wide range of people, some of them young, some more experienced, some think like him, others do not; and create a dynamic that leads to the right places.

Sometimes we deal with questions we have no answers to, problems, to which we see no solution from where we stand in time (unlike formal work procedures, wherein you know that if you take a certain path, you will get a certain, expected result). In these situations, it is necessary to create the process that leads to a successful solution. He is able to create a dynamic that eventually leads to results – a dynamic that combines professional and intellectual abilities, with an egoless ability to listen."

In point of fact, it can be said that what AA described were the fundamental skills of a leader.

If we were to present the central line of thought that stems from our interviews with the experts, it would be the statement that *systems engineering is management based on technological knowledge. It follows that a systems engineer is one who manages systems with technological infrastructures and must, therefore, be an engineer.*

This, however, is not a rigid pattern. Visualize a spectrum, at the one end of which is an engineer and at the other, a manager. On this spectrum, we can place systems engineers in various states of function. For instance, a complex project's lead systems engineer would be closer to the management end, while the systems engineer of a small, focused development team would stand closer to the engineering end. Yet, both would need to possess interpersonal and leadership skills, because both must lead people toward a common goal, while facing various constraints. *Dealing with people is one of the central properties that set systems engineers apart from other engineers.*

Be that as it may, it should be noted that this approach, though almost unchallenged by our interviewees, is not the prevalent opinion, and many systems engineers see the management of engineering processes as the center of their activity. The vast majority of them are, after all, engineers by basic training.

Prof. Olivier De Weck believes that the human factor in systems engineering is underappreciated. He agrees with Prof. Aviv Rosen's claim that among the reasons for this are the difficulty entailed in describing it in mathematical terms (a language engineers feel more comfortable with) and the challenge of overseeing it.

1.3.1 SYSTEMS ENGINEERING AND TECHNOLOGICAL PROJECT MANAGEMENT

The question of the link between systems engineering and management is mostly expressed in *the management of technological projects*, and the more complex they are, the more important it becomes. This is especially obvious in the intricate web of relations between the project manager and the chief systems engineer. *This pattern of relations has a substantial effect on the ability of these two to handle the ever-increasing variety of technological project complexities and to lead them toward successful resolutions.* This relationship structure is influenced by such components as the personalities and specializations of these two position holders, the experience they have accumulated, the organizational culture of the company tasked with the project, and more.

Below, we present a number of positions concerned with this important issue:

Dr. Eric Honour explains: "systems engineering and project management differ when it comes to priorities: project managers focus on the task, schedule, and budget; the technical manager (or chief systems engineer, or whatever you call him) is responsible for the results the tasks yield. The project manager wants to accomplish the mission; the systems engineer wants to see how well it has been accomplished. Their goals are identical, but their priorities are different. The project manager looks at the cost first, then at the schedule, and only then at the technical aspects. The priorities of the lead systems engineer are reversed."

Dr. Gillie Fortuna believes that the project manager is the chief systems engineer, even if, in the project, he operates alongside him. He explains his statement, by contending that in a complex technological project, "the project manager needs to have an engineering background. A project lead who is not an administrator, hires one to help him manage the budget, but he has to be a technological expert. This is why the project manager is the true chief systems engineer. The person defined as the project's chief systems engineer is a kind of deputy of the project manager, who also dedicates some of his time to administrative work."

Yossi Ackerman also finds no clear division of roles between the two: "The difference between a project manager and a systems engineer is insubstantial. A systems engineer sees the entire project from a technical-operational aspect. The project manager also sees the technical aspect, as well as other things, such as the economic and legal aspects. For all that, there is considerable of overlap between them. A systems engineer does many things the project manager does. In small projects, one person fills both positions. There is no structured framework defining the activity areas of each one."

Cecilia Haskins, who believes systems engineering and project management share a symbiotic relationship, summarizes: "In the past, no separation existed between these two fields. The managerial and technological components were handled together intuitively. But today, we get to a level of specialization so high that everyone is immersed in their own field and people become disjointed. Project management and systems engineering are like yin and yang (complementary opposites – a term taken from ancient Chinese philosophy – the authors) – one cannot succeed without the other."

1.4

THE EVOLUTION OF A SYSTEMS ENGINEER

One of the major expressions of systems engineering still coming into being as a profession is the fact that the main specialization in the field is based on *the accumulation of experience*. There are *graduate programs* in the field, but the training they offer strengthens and deepens the knowledge base of those who have gained experience in the systems fields. In most cases, companies recruit engineers who trained in the basic engineering fields. These engineers advance through the organization's career and job rotation paths. They are assigned to various professional and managerial positions, and some of them, those, whose fields of interests have moved beyond their basic professional area, become systems engineers.

All the systems engineers we have conversed with have experienced this type of development process[1].

Below are a number of examples, based on their own testimonies:

Sharon Shoshany Tavory: "At the Department of Digital Systems, where I worked, the prevalent perception was that real-time software development should be done by electronics engineers, because software people could not see the entire picture. They only saw their own 'bits'.

[1]Except for the special case of Eric Honour, who joined the US Navy, which has its own unique needs and qualities, and acquired an undergraduate degree in systems engineering within the framework of his service, as part of a combined training program for officers and system analysts.

Managing and Engineering Complex Technological Systems, First Edition.
Avigdor Zonnenshain and Shuki Stauber.

At the time, the term 'systems engineering' was already in existence, but it did not bear the meaning it does today. It was usually used to refer to someone who got promoted to Assistant Project Manager. In those days, I had begun doing systems engineering work. For example I, managed interfaces and analyzed processes, but I was not called a 'systems engineer.'

Every engineer starts his professional career as a 'screw'. He can choose to remain a screw, and focus on the field of engineering he specialized in, or he can choose to look around him. I chose to look around me."

Bud Lawson – a computer engineer by training: "As I look back and examine my work, I see that I have always practiced systems engineering. But that conceptual shift began, for me, in the mid 1970s, when I stopped focusing upon the computers themselves and started looking deeply into applications for computers. I felt that I was doing something different, but I did not use the word 'systems engineer.' I had worked at computer engineering previously, as well. But after I had started working on the applications, I told myself that I needed to learn much more about other areas, such as power distribution. If I wanted to design a suitable solution, then I could not focus just upon understanding the computerization part, I needed to understand the area that the computer was supposed to serve."

Mimi Timnat – a software engineer by training: "I joined meetings with clients, and was required to understand their expectations and look for solutions. I was exposed to engineering fields I had never dealt with, and terms I had never learned. At first, it was strange and unclear, but along the way, I asked questions, learned and began to understand many engineering subjects – far beyond mere software. I was naturally attracted to the need to understand the whole picture."

Henry Broodney – an electronics engineer by training: "In the beginning, I did not see myself as a systems engineer. Later, when I was integrated into more projects, I began to understand that I was not dealing with the electronics alone, but with the entire system."

1.4.1 THE MAIN PATHS OF DEVELOPMENT OF SYSTEMS ENGINEERS

There is, of course, no rigid pattern to systems engineering, and the discipline's work patterns are affected not only by the differences between organizations, but also by the differences between the personalities, personal preferences, and professional backgrounds of the field's practitioners themselves.

Nonetheless, it is possible to distinguish between two types of systems engineers. The first, we shall refer to as "Lateral, Management-Oriented Systems Engineers," and the second, we shall call "Professional-Disciplinary Systems Engineer." One of the marked differences between the two is in the answer to the question of whether systems engineering is a position or a profession.

"Management-oriented systems engineers" see it as a position. This means that as far as they are concerned, their function as systems engineers is part of their professional development. For instance, an electronics engineer, who becomes a systems engineer at a certain point, will stop being a systems engineer when he advances to the next step in his career and becomes, for example, a development or technological manager.

For example, Alon Gazit, who has never been called a systems engineer, agrees that at a certain point in his professional life, he had, in effect, been a systems engineer, while today, as a Senior Technology Manager, he does not see himself as one: "I am a development manager, who encounters many systemic dilemmas in his day-to-day work. Today, I no longer deal with specifications that need to meet client requirements."

We chose the example of a project systems engineer, because a substantial part of all systems engineers operate within project frameworks. There are, of course, also, systems engineers in what is referred to as "the engineering groups of the matrix" – the professional support groups that serve projects by dealing with technological developments.

Alon Gazit expands the distinction between the two types of systems engineers: "Management-oriented systems engineers are potential managers. They have leadership abilities and communication skills, and they are more willing to compromise. Suppose there is a multidisciplinary problem that needs solving, and it is not yet possible to even define where it originates, and, consequently, who should be handling it. This is the type of problem a good systems engineer needs to be able to solve, by combining leadership and analytical skills. In a case like this, a professional systems engineer would find the problem more difficult, because, although he has the necessary analytical skills, he will be limited by his lacking leadership ability. Professional systems engineers are more solid, more perfectionists. They enjoy dealing with technology, but not the small, technological details – technology in a wider sense."

Erez Heisdorf and Benjie Rom (respectively) further add: "Some systems engineers want to manage other people, to organize; they have the personalities of leaders. In contrast, there are those who wish to delve deeper into their fields and mature as professionals. They are not interested in managing people; they wish to focus on the technology.

Management-oriented systems engineers have the ability to take a broad view of the situation. They look at the product from the client's perspective as well. Compared to them, professional systems engineers certainly possess a broad perspective, as they must, but they focus more on the technical side, and less on the business aspects."

In technological companies, the importance of systems engineers is rising, and with it, raises the numbers of those who believe the systems engineer to be a critical position in the career of anyone who wishes to manage multidisciplinary systems.

Thus, for instance, at the conference of the Israeli Association for Systems Engineering, INCOSE_IL, which took place in Israel in 2013, The CEO of IAI declared that in his view, the company's systems engineers are a group of professionals, which, by its very nature, constitutes the company's executive reserve.

Because the systems engineer position is still amorphous, many organizations have yet to devise designated promotion paths for it. Systems engineers advance in an organization by being assigned jobs that entail responsibilities over systems with ever-widening scopes; they can be promoted to the position of chief systems engineer, or to project management positions, which require an in-depth understanding of systems and a well-established technical background.

In many cases, a separation between the career paths of project managers and those of systems engineers does exist, but it is not absolute separation. In technological organizations, such as those engaged in defense systems, aviation, or space, many projects managers grow out of the chief systems engineer discipline. Contrastingly, in business organizations, project managers may evolve out of marketing specialists and business managers.

1.4.2 THE EVOLUTION OF SOFTWARE ENGINEERS INTO SYSTEMS ENGINEERS

Over the years, an ambivalent connection has existed between software engineering and systems engineering. On the one hand, systems engineering is a methodological tool meant to support the design of engineering systems and integrate between them; a systems engineer needs to understand and show interest in the engineering fields that comprise the system he is in charge of. Software engineers, however, have often been perceived as those who concern themselves only with software and have no real knowledge or understanding of the world of classical, physical engineering. On the other hand, there are many similarities between systems engineering and software engineering, because both deal with abstract systems. Both systems and software engineering require virtual models in order to perform at least one of their common, overarching missions, namely: the integration of technological systems.

In the words of Sharon Shoshany Tavory: "In systems engineering, there is a lot of abstract discussion that needs to take place before moving on to the 'physical' stages. It is different from classical engineering fields. In many cases, engineers who focus on the physical aspect and attach little importance to the abstract aspect are criticized. Software engineering, however, is abstract in its very essence. Software specialists do not produce a physical product. The blueprint of the product – the code – is the product. And so, in time, the work methods of systems engineers have become more and more similar to those of software specialists."

Conversations we held in this context have raised, on the one hand, arguments, according to which software specialists tend to concern themselves with software and are less willing to deal with the other engineering disciplines. On the other hand, some argued that, because of the increasing importance of software in the creation of systems and supersystems, more and more systems engineers rise from the ranks of software engineers.

Mimi Timnat: "Alongside those who prefer to specialize in software, there are software people with a tendency towards working with systems, who enjoy working in large projects, in systems of systems, and being able to understand the whole picture. These people choose to evolve into systems engineers, and thus, become familiar with other disciplines. Furthermore, it is important for the chief systems engineer to have a good grasp of the dominant area in the project he is developing. In software-heavy systems, software engineers have a natural advantage."

This also explains the increase in the number of software engineers who become systems engineers:

Mini Timnat: "In the past, hardware components had much more dominance in systems than software components, and so a considerable part of the systems engineers rose from disciplines of relatively high technological complexity, such as electronics. Today, software takes up much more weight, causing more systems engineers evolve from that area."

Prof. Aviv Rosen is of the opinion that the major share of software engineers still prefer to focus on software, and only a minority is willing to tackle other fields. He believes the reason for the growing number of systems engineers who rise from among the ranks of this group of engineers is that software engineers have greatly increased in numbers in recent years (in many projects, they constitute the largest group of engineers). He claims that if we look at the percentages, we will see that the relative share of systems engineers who started out as software engineers is still fairly small.

1.4.3 THE TRAINING OF SYSTEMS ENGINEERS

It is commonly assumed that *systems engineering is meant, first and foremost, to serve the needs of the industry*, being a profession that emerged and evolved, mainly as a result of needs in the field. The head of the systems engineering program at the Technion, Prof. Aviv Rosen, says that this need stemmed from the fact that more and more large systems were failing. They were either technologically or economically unsuccessful, or simply failed to meet their deadlines. Organizations, mostly in the technologically and systemically complex fields of aviation and defense, recognized the problem and began to develop their own training programs, some of which were very extensive and numbered hundreds of hours of study.

Prof. Aviv Rosen: "In Israel, this trend picked up speed during the nineties, mainly in the defense and aviation industries, which felt that systems engineers who acted as such on their own accord were not enough; they needed to be given tools and methodologies to help them bring order to their applications of systems engineering."

At a later stage, the industries approached educational institutions that taught engineering and asked them to develop training programs that granted academic degrees in systems engineering. Run in collaboration with industry officials who even take part in the academic steering committees, these programs are mostly intended for engineers, and their graduates receive Master's degrees.

This brings us to a question: why would an institution that mostly concerns itself with research want to provide professional training that largely deals with practical applications? To this, Prof. Aviv Rosen replies: "The ultimate goal of engineering schools is to train engineers. An institution that performs research can train engineers better than one that occupies itself only with training." Also, in the context of Israel and the Technion, he adds: "The founding fathers of the Technion had established that one of its goals was to contribute to the needs of the State of Israel; training a systems engineer, who would benefit Israeli industry, is one way of realizing this vision."

As the CEO of one such industry, Yossi Ackerman from Elbit relates his angle: "A good systems engineer can manage without furthering his studies. This is true for every field: there are exceptional teachers who have never studied pedagogy, and there are, of course, those who have. There are excellent systems engineers who never attended formal training frameworks. They possess the right qualities and are self-taught.

Nonetheless, continuing education programs have an added value. They have put things in order. They poured meaning into what systems engineers were doing. These programs also create a common denominator among Elbit's systems engineers, each of whom had arrived from a different unit. As time passed, because of the growing importance of the systems engineer's role in each project, we wanted to institutionalize the field and asked the Technion to create better-founded educational frameworks that awarded Master's degrees."

Not everyone believes systems engineering training can only take place in academic, Masters' degree programs, meant for experienced students. One such contender is Eric Honour, who, based on his unique experience, does not dismiss the viability of basic systems engineering studies: "There are undergraduate programs available in the US, and I believe it is entirely possible to study this field without prior experience. After graduation, one can gradually be integrated into the practical field. It is not very different from an electronics engineer who studies for 4 years and then starts working at an organization, having no prior experience. At first, he is first attached to a mentor and charged with a relatively simple task. Similarly, a newly graduated systems engineer can first be assigned less complex tasks and aided by a mentor."

Being also the owner of a company that offers systems engineering training courses, Eric Honour presents the differences between companies like his and academic institutions: "Many of the lecturers in these programs are academics with no hands-on field experience. There is no competition between these studies and the ones provided by training companies; the two complement each other. However, in recent years, in order to generate more income, universities have begun to offer on client-site short term courses. For this purpose, they also offer instructors with field experience. This particular activity indeed competes with private training companies."

1.5

SYSTEMS ENGINEERING IN VARIOUS ORGANIZATIONS

It is apparent that systems engineering wishes to provide solutions, mostly in innovative, complex systems that require orderly work processes, to minimize the systems' development and implementation risks, reduce the risk of possible failures, and have ways of handling them successfully, if they do happen. In addition, systems engineering deals with changes in the demands of the market and the needs of the clients, as well as the technological changes that affect systemic solutions. These are the underlying reasons for some of the differences in the extent to which systems engineering is implemented in various industries.

Systems engineering is particularly evolved in the aeronautics, space, and defense industries, because the engineers of these industries have been tackling complex projects as early as the 1940s (the Manhattan Project) and 1960s (the Apollo Program). These projects required extremely high system reliability and safety levels. Furthermore, these projects faced very tight, challenging schedules. Compared to them, other branches of industry are still in the process of carefully examining systems engineering, weighing cost/efficiency considerations, and gradually adopting it into their work.

Naturally, much of the defense industry (and of the aeronautics and space industry as well) is government funded and operates as part of the public sector. Prof. Olivier De Weck finds substantial differences between the business and public sectors, in terms of their willingness to adopt systems engineering work patterns: "Systems engineering in the public sector, in government or defense projects (which usually

Managing and Engineering Complex Technological Systems, First Edition.
Avigdor Zonnenshain and Shuki Stauber.

are also government projects – the authors), is integrated into the system, an inseparable part of the requirements specification. The business sector, on the other hand, is focused on immediate or short-term benefits, and so, only uses systems engineering methodologies if it has added value, namely, financial profitability."

Below are a number of examples that demonstrate the difference in the implementation of systems engineering by various industries:

Boaz Dovrin, a systems engineer who transitioned from a company in the defense industry (Elbit), where systems engineering had been deeply rooted, into a company in the field of medical equipment (Phillips Medical Systems Israel), at a time when it had just begin to incorporate systems engineering work methods among its employees: "I understood from the questions they had asked me (in the job interview) that they did not know what systems engineering was. They were basic questions, completely out of place for someone who had arrived from Elbit ... The gaps between Phillips and Elbit were so large that I could not understand how their projects worked, how they were able to manage multiple projects without synchronizing their resources."

A similar testimony, by Benjie Rom, who also transitioned from The Elbit Group (namely, its subsidiary, Elop) into digital printing equipment company, Indigo: "At Elop, the systems engineer is responsible for designing a part of the system, while at Indigo, he has no part in the design. Here, a systems engineer can share his experience with the designers, or take the group in a certain direction, but the planning itself is done by the matrix bodies. The main reason for this is the complexity of Indigo's products, which necessitates the placement of a systems engineer in each technological group, thus reducing the need for the systems engineers to deal with the project's more technological components."

Further explanation is offered by one, who transitioned from the defense industry to the chemical industry – Gillie Fortuna – appointed to the position of CEO of ICL-group's TAMI Institute for Research and Development: "Systems engineering is important in a system with numerous components that require trade-offs to be made. In the chemicals industry, most of the systemic view stems from the need for optimization between several products, some of which are beneficial, while others are attached as part of the process. There are not as many alternatives as there are in aeronautical systems. It is possible to examine alternatives, considering the purity of the material and the cost of the product, but it does not compare to the complexity and high level of the alternative examination process required to launch complex airborne systems into the air. In the chemicals industry, the final test is the application of the development to competitive, economic production. In the end, this necessitates a systemic view of all the development and economic production capabilities, but at a lower level of complexity."

In comparison, the oil and gas industry is characterized by significant safety-related constraints and complex systems, and still, it has yet to successfully integrate the discipline into its work patterns.

Prof. Olivier De Weck demonstrates: "A lot of offshore oil drilling takes place in shallows, but major incidents happen in deep waters. These drilling projects are complex systems that have to handle extreme conditions, not unlike those of space exploration, namely, working with robots under high pressure, at high temperatures, and at distant locations. In spite of all that, when asked about systems engineering, the people of this industry usually respond by asking what that is. The first signs of the implementation of systems engineering are beginning to emerge in refineries founded today, but things are still done sloppily, and the dangers are many. When the system operates at low temperatures and there is a leak, the leak is repaired and the problem is resolved. But when the pressure and temperatures are high, the same leak becomes a serious problem."

Dr. Cecilia Haskins also mentions this industry: "For products located in extreme environmental conditions there are many challenges, both technological and physical. The heads of the oil and gas industry have only recently begun to recognize the fact that systems engineering can help them find solutions to some of those problems."

In this context, Prof. Olivier De Weck says: "The problem with systems engineering in the business world is that its short-term benefits are somewhat hidden. Even if great efforts were invested into systems engineering, the benefits will only emerge after a period of time, which could be several months or years. When a complex system lasts many years, people will talk about what an impressive job the systems engineers had done on it, and how they should be thanked and appreciated for it. But after so many years, those systems engineers will not receive the recognition they deserve, because by then they will have retired or moved away. The gap between cause and effect here is very wide."

The differences in "systemic" work patterns are not limited to those between different industries. Each company has its own, unique organizational culture, and it affects the systems engineering work patterns of that company. The basic work methods may not vary greatly, but the ways they are implemented and the placement of the emphases, change from company to company, even within the same branch of industry.

For example, Dr. Kobi Reiner distinguishes between companies within the defense industry that began as defense R&D units and companies that began as production and maintenance arrays. Companies of the former category emphasize the first stages of a project, using numerous systems engineers during the development phase, while companies of the latter type emphasize the more advanced stages – the ones that deal with integration. Naturally, these differences influence the types of systems engineers working at both kinds of organizations, the manner of their work, and the methodologies they employ.

1.5.1 WHO IS A SYSTEMS ENGINEER? – A QUESTION OF TERMINOLOGY

Another way, in which organizations differ from one another, is terminology. A systems engineer can be known by different names. In Elbit Systems, for instance, a project's lead systems engineer is called a "technical manager." Yossi Ackerman, who, until recently presided over Elbit, says that there is no exact definition for a good systems engineer, and that is a good thing: "The position needs to be given space, and then defined in accordance with the given situation. A manager, who is an engineer by profession, and who looks at the whole technical and technological picture, can be called a systems engineer."

The opposite is also true: in some organizations, those referred to as "systems engineers" do not actually practice systems engineering. An example of this is computing giant IBM, which, according to Henry Broodney, employs thousands of so-called "systems engineers," the vast majority of whom do not fall under the commonly accepted definition of a systems engineer at all. Rather, *they are Information Technology engineers, who use systems engineering methodologies in their work* – systems engineers within the IT field.

Moreover, unlike most industries where the systems engineering discipline is well rooted, most of the systems engineers at IBM are not in the company's research and development bodies, but *in the sales, marketing, and service divisions*. This way, they speak the same language as the systems engineers employed by potential clients.

1.6

THE FUTURE OF SYSTEMS ENGINEERING

In the world of technology:

It is clear that the prevalence of the main trends that had created the need for systems engineering methodologies (the increase in system complexity, time and budget constraints, and the extent of the clients' involvement) will only increase, and so will the need to systems engineering, mostly in technological organizations.

Today, systems engineering is mostly implemented in defense, aviation, and space enterprises. The estimate is that organizations in other fields – such as communications, medicine, digital printing and photography, or smart transportation systems – will adopt the same systems engineering processes that the defense and aviation industries have found to be so effective. Evidence of this is the rising number of students from other industry fields to join the systems engineering studying programs.

The emergence of systems engineering in these organizations is backed by the ongoing process of the field's institutionalization. Academic institutions are opening training programs for systems engineers. One such institution is the Technion, the only university in Israel to offer such a program: so far, its master's degree program has trained approximately 900 systems engineers. Israeli colleges are also offering academic programs in systems engineering.

Having come to understand the importance of systems engineering, more and more technological organizations are willing to integrate its work methods into their systems and even train systems engineers themselves. The managers of these organizations are also coming to the realization that systems engineering is the all-important

Managing and Engineering Complex Technological Systems, First Edition.
Avigdor Zonnenshain and Shuki Stauber.

link that connects them to the technological specialists, who develop the products of their organizations.

For example, at the INCOSE_IL (The Israeli Association for Systems Engineering) leadership panel, executives saw systems engineering as a discipline that includes tools and methodologies that help them make decisions throughout the lifetime of a project. They also saw it as a discipline that facilitates the creation of a synergy between engineers from various fields and thus helps put the project on the right track toward success.

Outside the world of technology:

It is currently impossible to estimate whether the evolution and growth of systems engineering will continue mostly within the boundaries of the technological world or penetrate other fields as well. This is not merely a practical issue; it is also a question of image and branding. On the one hand, systems engineering has emerged from out of the technological world, with the purpose of answering its needs; and the very fact that it is called "engineering" is suggestive of its native path of development: to remain, for the most part, within the technological world, as *engineering with management aspects*.

On the other hand, it may be that, in time, as other areas of practice gradually discover its benefits, nonengineers will also seek to adopt its methods, in their non-technological content worlds. Thereupon, within these circles, systems engineering will become *management with engineering aspects*. It is also possible that these content worlds will expand beyond the spaces of management and engineering, supported by the tight connection between systems engineering and systemic-holistic thinking. Then, it can be implemented even in such areas as education, human resource development, politics, and more.

This approach may gain momentum due to the increasing importance of planning, a central pillar of systems engineering: Kobi Reiner says that, in situations of budget and other constraints, advance planning is of crucial importance, as it allows one to reduce the risk of failures. As Prof. Ovadia Harari stresses the decisive influence early-stage planning processes have on a project's chances of success. Norman Augustine completes the picture, by saying that methodologically, budget planning is no different from engineering planning.

If these approaches become prevalent, systems engineering will find a home outside the technological world too and will be adopted by an ever-widening range of areas of activity.

Additional trends:

The rise in the importance of the human factor in systems engineering is unmistakable. The leaders of the field, the vast majority of whom are, as we remember, engineers, place more and more weight on the importance of the systems engineer's interpersonal skills, in his position as the leader of multidisciplinary teams of engineers. At the same time, there is a rise in the importance of such qualities as simplification abilities, helpful when designing complex systems, which are naturally more prone to human error.

The importance of the human factor is also expressed in the realization that the user of a system is a central component within it. This is why today, there is an increasing trend of designing systems that reduce the chance of the user making mistakes in their operation, due to their structure and complexity. Another way to present it is this: in the past, user failure was considered human error (ergo, the user's fault); today, the realization that these errors can be caused by the structure of the system (ergo, the system's fault) is becoming more common, and so they are called "usage failures."

Future developments in the computing fields will also affect the computerization of systems engineering processes. Advanced computing systems have been in the service of traditional systems engineering for years; yet, only in recent years have they begun to serve systems engineering as well. These newly developed computing systems are based on the clever use of models and simulations, successfully implemented in traditional engineering fields.

In systems engineering, this new combination should instigate changes in the traditional development process (the V model). Use of models and simulations in the development phase (the downward slope of the V model) allows the validation of system behavior models, using simulations early in the development process, thus eliminating the need for the construction of complex and costly physical testing systems.

Yossi Ackerman mentions the Internet's impact on systems engineering. According to him, the exposure of the public to large volumes of information has brought on a devaluation of the all-knowing experts and an increase in the importance of interdisciplinary teams. In a complex world, only experts from different fields thinking together can produce added value, and interdisciplinary team management is one of the fundamental abilities of systems engineers.

PART II

A WORLD OF COMPLEX PROJECTS – THEN AND NOW

2.1

THE IAI LAVI PROJECT – THE DREAM AND DOWNFALL

An interview with Prof. Ovadia Harari of Blessed Memory

During the second half of the 1970s, the Israeli Air Force had had four main types of attack aircraft: two single-engine fighters, the Skyhawk and the Delta; the Shachak, Nesher, and Kfir models; the two-engine Phantom; and their leading model, the two-engine F15. Around that time, the Air Force began to see the need to replace the Skyhawks and the Deltas with a more advanced, single-engine aircraft.

Israel Aerospace Industries, the company in charge of producing the Nesher and Kfir models, understood that the day when these types of aircraft would no longer be in demand was not far off and was preparing to face the task of developing a new fighter plane.

These preparations were not being made in response to a specific request made by a client. The discourse concerning the need for a new aircraft never went beyond the spontaneous, informal expression of ideas and standpoints between high-ranking Air Force officers. Naturally, these officers maintained continuous contact with IAI, and so, inevitably, word traveled. In other words, because the Air Force had not presented any formal request or demand for the production of such an aircraft, it was impossible to tell what sort of performance it was required to deliver. Moreover, whether the need would even be realized remained unclear. For instance, the Israeli government might

Managing and Engineering Complex Technological Systems, First Edition.
Avigdor Zonnenshain and Shuki Stauber.

not have approved an order for a new plane, be it from an Israeli manufacturer or a foreign one.

2.1.1 THE FEASIBILITY STUDY

Despite the uncertainty, IAI was prepared to take a calculated risk. The company wanted to be ready for the (in its estimate, fairly likely) scenario where a demand for the development and production of a new aircraft would indeed be made. At the same time, it wanted to prove that it was capable of accomplishing such a task. So, the management of IAI decided to perform a feasibility study of the new aircraft's design and production.

The task of performing the study was placed on the Department of Preliminary Planning, a part of the Engineering Division. In the words of Ovadia Harari, who then headed the department, its job was "to sprout new ideas and examine solutions without there being anyone to place an order for the product."

But how can one provide solutions, if one does not know the problems?

Harari: "We defined the needs ourselves. We prepared a requirements document based on two main sources of information: first, we held discussions with Air Force officers. This allowed us to assess their needs in relation to the new aircraft. Second, we, as a professional department, were familiar with the capabilities of the aircrafts that existed at the time. We knew the performance they could deliver, and which technological elements we would need to use to improve on that performance. We reverse engineered existing aircraft models (reverse engineering is taking an existing product apart to study its inner workings and to figure out how to make similar products – the authors)."

A feasibility study is not a carefully planned, well-organized process, because there is no client to define the needs and requirements. There is therefore no strict timetable and essentially no budget constraints. The role of "client" was filled by the IAI management, whose only demand was that the department estimate what the potential user might want.

When a client is present, this task naturally falls under the responsibility of his people. Had this been the case, the Air Force project manager would have prepared a base document and distributed it among the relevant officers within the Air Force. Each relevant position holder would then have expressed his opinion on the matter, and eventually, they would all have met and discussed it. The project manager would have served as a mediator and coordinator between the Air Force's various professional units, and the process would have resulted in a balanced requirements document submitted to the company undertaking the project.

Harari: "Our client did not speak with one, cohesive voice (because the Air Force had not yet formulated a concrete position on the matter), so we had to present the management of IAI with the full range of opinions we had heard from the Air Force. For example, one specialist officer believed the aircraft should have a certain flight range, while another was willing to settle for a shorter one, so long as the plane had a more accurate arming system."

The information gathered by the feasibility study team (a team of 5, led by Harari) allowed it to take the first step in the process of defining the new aircraft's configuration, namely, its size.

The size of an aircraft is a very important base parameter. The size affects the plane's capabilities (such as speed and flight range). The type and nature of such subsystems as the engine, radar, flight control, and electronic warfare (EW) are also derived from the size. Of course, the size also has a direct effect on the total cost of the project.

In the beginning of the process, various configurations were considered, including both single and two engine options. The cost was the main reason for the decision not to make a two-engine aircraft (a Lion) to replace the F15. Producing a two-engine fighter plane is a mission of enormous proportions, its costs so high that only a handful of countries in the whole world are able to take it up (all of them, obviously, much larger than Israel).

The need to determine the types of the plane's subsystems drove the team to meet with various experts in IAI. These were leading experts in the professional departments that specialized in each of the relevant systems.

The Political Aspect

As the team was preparing the future aircraft's technical specification, the management of IAI was hard at work on the political level, attempting to urge the Air Force to place an order for the project with IAI. This was a complex task, because whether or not the order was placed did not depend solely on the potential client – the Air Force – but on two other important factors: the Israeli government and the US Department of Defense.

Ovadia Harari, who had later become the right-hand man of IAI's CEO, explains the situation: "The cost of that development project could easily have reached two billion NIS or more (with over 200 planes acquired, the sum would have climbed up to around 20 billion NIS). All projects that approach the cost of one billion NIS immediately gain a political aspect, because they require considerable resource allocations from the state budget. In situations like these, the question of 'how much will this cost' is not the only one asked. It is accompanied by such questions as 'how many people will this provide employment for.'

On top of that, the politicians needed to talk to the US Department of Defense, because Israel cannot develop aircraft engines – the cost of such projects is beyond its financial reach; there can never be an aircraft of pure Israeli origin. So, to design the new aircraft, we needed to receive information about engines, which required the approval of the US authorities. The US manufacturers of aircraft engines, General Electric and Pratt & Whitney, do not give out attack aircraft engine specifications without the approval of the Department of Defense. Had IAI made the request without such an approval, the Department of Defense would have launched a painfully thorough investigation and probably refused us outright. The company needed political support and backing. Otherwise, why would the Americans help IAI? After all, it would be more profitable for them if the Israeli Ministry of Defense bought the plane from an American company."

Back then, over 30 years before these lines were written, the question of cost in military projects was nowhere near as important as it is today. At the time, projects were priced using the "cost plus" method. This meant that the client paid for the development, however much it ended up costing. No finite budget was determined. Nevertheless, the financial question weighed heavily even on the members of the feasibility study team, due to the project's extraordinarily high cost prediction.

Harari: "The very fact that our goal was to develop a plane to replace the Kfir and the Skyhawk and not the F15 stemmed from the cost issue. Meeting the various needs of the military is much easier with a large aircraft. In aeronautics, you control the money through size. The cost of a plane is a function of its size and weight."

The technical specification for the Lavi fighter aircraft was slowly coming into being, taking into account the various needs and constraints, including the cost limitations. Beyond the definition of the plane's configuration and capabilities (flight distance, load capacity, etc.), systems were adjusted using trade-offs.

Explanation: one of the main objectives of systems engineering is to optimize systems by making trade-offs. For example, it is possible to install a more powerful engine, but then the plane would need to be made larger; or, the radar system could be upgraded to a more advanced one, but the new system would weigh more, which would put a heavier load on the aircraft.

Several months after the process had been initiated, a breakthrough occurred, which dramatically raised the chances of the project being realized by IAI.

Ovadia Harari: "The Minister of Defense (along with the IAF Commander) decided to visit IAI, and we presented the project to him. Although I was not a management member at the time, I attended the meeting as an expert, there to answer professional questions. Note that both Defense Ministers who had been involved with the project had some knowledge of the area (Ezer Weizmann, who was in office when the project was announced, was himself a former IAF Commander; while Moshe Arens, an aeronautical engineer and a professor at the Technion, had served as Vice President of IAI in the past – the authors). At the end of the meeting, Ezer Weizmann said the words that made our day, or rather, our year: 'I find this project interesting. I will handle the political aspect and talk to the US Defense Minister.' We felt that we had gone up to a new level, for without that support, our chances would have been slim.

The cost assessment had not been presented at that meeting, because we had not yet had enough data to perform a budget evaluation. Only later, in 1979, did we provide the Ministry of Defense with the assessment that the development would cost 780 million NIS.

However, to all those involved, it was clear that this was only a base figure; meaning that the actual number could only be higher."

One might wonder why the client should care for the development and its cost. Should not the client, who is only interested in new planes, be told only the price he would have to pay for one aircraft or for a given number of units?

Harari: "This is true for civilian projects. If a company wants to develop a new X-ray scanner or a new car, it must indeed bear the risk. Not so in military projects, where the sums invested in development are so high, that the industry cannot afford to take the risk, unless it is guaranteed to have a client for the product. So, the client would give the company a guarantee early on, during the development phase. Today, things are different. The military client still invests in the development, but he expects the company to partake in the risk."

Harari believes that even the work of the team that performed the feasibility study (a task that, as we said before, lacked organized structure by definition) included many systems engineering elements. This is due to the fact that it had major "systemic" properties, such as the integration of different fields, the need for a broad perspective, the weighing of alternatives, the characterization of the needs of a client (even if it was only a potential client), etc.: "This is the job of a systems engineer. He should be able to handle preliminary planning and system architecture design, to lay out the general lines of the solution – he is the manager of the system architecture. However, we must not forget we are speaking of a time when systems engineering was just beginning to take its first steps. The very term 'systems engineering' had not even existed yet. Systems engineering existed mostly in people's minds, which is exactly the problem it is trying to solve: the goal of systems engineering is to get these ideas out of people's minds and into the field, to encourage them to act based on structured methodologies that can be tracked, rather than have each person follow his own instincts."

After the order for was received from the Ministry of Defense, the newly initiated project began to take shape and rely on orderly, systematic activity. But, Harari admits, even then, it was not the carefully organized project one would expect to see from today's systems engineers: "30 years ago, there was no orderly paperwork and work processes were not systematic, like they are today. Thus, for instance, you would not find an archive with SDR (System Design Review) documents dating back to that time, because such things did not exist, then."

2.1.2 THE PROJECT

The two teams, technical and political, pushed on in earnest. The feasibility of the project was ripening: "The technical specification was expanding, a more detailed budget plan was taking shape, the Americans agreed to provide us with engine specifications and the Minister of Finance was willing to allocate resources for the project. Even the Prime Minister was involved (Ovadia Harari tells us that Prime Minister Menachem Begin, in his picturesque way of speaking, said: 'when the IAF commander asks, I stand at attention'), and in February of 1980, the decision was made to launch the project.

Now we had a client – the Ministry of Defense – and an end user – the Air Force – both of whom are very skilled at presenting demands."

IAI appointed a project directorate, led by Ovadia Harari. The team was comprised of the same group that had performed the feasibility study, joined with other professionals. In the early stages of the project, the team numbered 10–15 members.

The Preliminary Planning Stage

The team began to write requirements documents, which were then given to the company's professional departments. For example, the department of aerodynamics was given a document that detailed the aircraft's configuration requirements.

Harari: "We asked the department's personnel to present us with variations on the given configuration, such as a slightly larger wing or a yoke of a certain size. We also asked them to analyze possible limitations and test them in a wind tunnel (a wind tunnel simulates the flow of air around an aircraft in flight and allows the testers to measure the forces and momentums applied to the aircraft – the authors)."

This was how the matrix management system worked. The project team directed the employees of the professional units by way of the requirements documents. This way, a chain of client–supplier relations was created: just as the Ministry of Defense was the client of the IAI project directorate, the project directorate was the client of the professional units, where dozens of engineers dedicated an ever-increasing amount of time to fulfilling the requirements they had been presented with.

Recruitment for the Lavi Project

The process of recruiting the project's preliminary team was based on the principle of bringing together knowledgeable and experienced people from different professional units. One does not recruit new employees from outside the organization for a project of this scope and complexity. The project's employees must be experienced, skilled, and intimately familiar with the inner workings of IAI, as well as with their colleagues.

Harari: "I consulted my boss and other people in the company, and together we defined whom the project needed; for instance, someone who knows systems, a good radar expert, someone who is well versed in advanced flight control (with the new aircraft, we were entering the electronic flight control category), someone who knows about aircraft structure, someone who specializes in aircraft aerodynamics and so on."

Of course, recruiting experts from the various specialized departments was no easy task. Even knowing the importance of the project, no department manager was happy to give up his best men. As far as they were concerned, our recruitment effort set back the work of their departments. From the outset of the project, the workload of the department managers and their remaining personnel was increased considerably. Moreover, to fill the void, he had to recruit new employees from outside the organization and train them; a process that took time and management resources. In other worlds, the specialized department manager had to cope with the organizational shock suffered by his department, all for someone else's project.

To deal with these constraints, the project administration cooperated with high-ranking executives in IAI; those in charge of the management of the project and the respective specialized departments.

A look at the types of people Harari chose for the project shows that their qualities are what we now know as the qualities that make a good systems engineer.

Harari: "Systems engineers are prominent team members. They are dynamic individuals, involved in many different areas; they do not shy away from anything.

They are open-minded; they ask questions and have a dialogue with you (as a manager). If you want something from them, they don't just go and do it, they ask you what you need it for. They sit with you and examine whether a different solution is also acceptable, they test the extent of leeway you give them. Over the years, I have worked with many such individuals in the company. I had gotten to know them. And when the project was underway, I came to my boss and told him whom I wanted. I wanted this person from Electricity department, that person from the Radar department and a third person from Flight Control department."

On Managing People

With great sincerity, Harari tells us of the way he had managed his team, and the lessons he learned, mostly in relation to his centralized management method.

When asked whether he consulted each expert on questions related to his own field, or generally asked everyone's opinion on everything, he responds: "Today I know that it is better to do all the work together. Not working this way (and many project managers still do not know how to do this right, even today) produces inferior results. People need to be included.

I was a centralist. I did not include them enough. Still, there were those among them who would not let me do what I wanted. Looking back, I now see that I had acted incorrectly."

But why choose this way over others?

Harari: "Today's management approaches encourage sharing. But back then, the manager was perceived as a macho, and his word was law. Managers thought they knew everything. Moreover, at the time, managers were very dominant, and so they found it easier to manage this way. There were very few objections, because managers were respected. Thus, my attitude did not cause any arguments to speak of, because people appreciated me.

This dominance became especially prominent when deadlines began to loom overhead, and the manager was under pressure to quickly move the project forward. In these situations, people needed to be spurred to work harder, which encouraged machismo. But this was not the right way to manage. In time, I learned that good systems engineering is cooperative systems engineering."

That being said, the dilemma of whether or not to share is quite real: if one of the main constraints many projects face is indeed tight schedules, and consulting and sharing take up valuable time, might it not be better to share less and centralize more, if only for practical reasons?

Harari believes that, in these cases, the downsides of a centralized approach outweigh its benefits.

Harari: "At the end of the day, it is not more practical. The sharing approach does take more time (in the short term), but it increases people's sense of responsibility, because they feel involved. If you just order them around, they become small minded. Systems engineering is people-oriented in essence ("people-oriented" is a term taken from a management model that places managers on a scale: at one end, stands a "people-oriented" manager, while at the other stands a "task-oriented" manager – the authors). If you cannot share and seek advice, you have failed as a systems engineer. You must convince your employee that your way is the right way. You have to compromise, otherwise people become small-minded. When you do not let people express themselves, they close up. Even if they have good ideas, they do not express them. They say to themselves: 'this manager has already made his decision; he doesn't want to be confused with facts.' If you make all the decisions yourself, you may save time in the short term, but if you make the wrong decision now because you failed to include other people, and six months from now, that mistake causes a problem, it will be very hard to correct it. Moreover, the correction will take a very long time, because you will have to go back to square one. So, when you look at the bottom line, which approach is more time consuming?"

The Structure of the Project's Directorate

The centralized management method also dictates the structure of the project's management. Harari addresses two important lessons he had learned in this context:

The first lesson concerns the fact that Harari effectively filled two positions: that of the project manager and that of the chief systems engineer.

Harari: "The fact that I had to fill both positions was detrimental to the project. I thought I knew everything and believed myself able to work 18 hours a day.

The lesson I learned was that in a project with a budget scope of 10–15 million dollars (and a team larger than 10–15 people), the two positions of project manager and chief systems engineer must be separate. In a project of this magnitude, there needs to be someone whose job entails mostly management – the management tasks take up most of the project manager's time. If he fills both positions and has to handle the technical issues as chief systems engineer on top of his management responsibilities, processes are delayed or produce mediocre results. He is forced to make decisions quickly and has no time to consider all the aspects and implications. Alternatively, decisions are postponed, which ramps up the costs.

In these, large-scoped projects, a chief systems engineer must be the deputy of the project manager. He must handle the project's professional aspects like electrical systems, software, or electronic warfare. He should not have to deal with the management aspects, unless they are entwined in the technical work. For example, he should be included in the project manager's meetings with the client, in order to present the technical aspects.

In smaller projects, one person can fill both positions."

Discussion:

The dilemma of whether or not to separate the two positions raises a principle question about the interface between systems engineering and management. The defining traits of systems engineering, such as the ability to integrate and coordinate various elements, or the ability to see things from a systemic perspective, can also be found in management. The question, therefore, is: to what extent is the practice of systems engineering also the practice of management, and is the separation between these two fields not artificial?

Harari's words suggest that systems engineering is a management practice based on technological knowledge. Technology, by definition, always entails the integration of subsystems. It follows that if a manger is one who manages organizational systems, then a systems engineer is one who manages technological systems. He, therefore, has to be an engineer by trade, because he has to understand the field he manages. Thus, unlike the professional manager, who is able to move from a management position in one area to a different management position in an entirely different area for "management is management," a systems engineer can only be effective when dealing with his own technological specialization. To conclude, Harari suggests that a system engineer is an engineer with management skills that allow him to perform the necessary technological integration, while using abilities that lie outside his areas of technical expertise (such as budget management).

The second lesson concerns the size of the project team and its effect on the work pattern with the company's professional departments.

Harari: "The Lavi project team increased in size as the project progressed. Eventually, the number of people approached 100. Each system had its own systems engineer, who was in charge of advancing its development and working with the professional departments. There were also systems engineers in charge of the development of subsystems (such as the flight control computer).

This was not the correct approach, because it made the team too dominant. There were so many good people in it that it ended up overshadowing the professional units, which created many disputes. The team dictated the solutions to the units and was too involved in their work. Professional departments do not like this level of interference. It is in their nature to say: 'you have given me the specifications for what you want, now step aside and let me do my job.'

A situation was created, where a powerful group that controlled the money and the client was making the lives of the specialized department teams miserable, causing them to become small-minded. The specialists thought: 'even if I have other ideas, the project administration will never let me express them, let alone try them out.'"

Later, Harari implemented this lesson in other projects he managed and delegated some of the authority to the departments that provided him with their professional services: "Instead of the systems engineer in charge of the navigation system being a member of the project team, he was a member of the professional department. In the next project I managed, which was no smaller in scope than the Lavi project, the project team numbered only 20 members. The systems engineers in charge of developing the aircraft's systems were members of the professional departments. In areas pertaining to the project, they were subjected to me; but they remained within their respective systems and the department manager continued to serve as their direct supervisor. This approach made the department more committed to the project."

To counter our argument that this approach forces the project manager to give up some of his power, Harari says: "This is the right way, so you have to compromise: you give up some of your power, but you get dedication in return. This is how I tie the departments to the project. Had I not done this, I would have had 90 commandoes surrounded by servants – that is no way to work. The right way is to form a small commando force and surround it with people who are committed, and feel that they are half-commando themselves."

Eventually, after about 2 years of preliminary planning, the concrete requirements for the Lavi plane were finalized, and the configuration and structure of its systems were defined.

Ovadia Harari expands and demonstrates: "By then, we had obviously already decided which engine the aircraft should have. We knew the size and weight of the radar system; we knew what performance to expect from it, how it would interface with the other systems on the plane and how much it would cost. The same was true for the fuel system: we already knew how the fuel tanks would be deployed inside the plane, how the fuel would be transported from place to place, how the tanks would be filled and emptied, how the aircraft would suction fuel even when flying upside-down, what the fuel flow rate would be, and which pumps we had to use to achieve it."

Costs and Budgeting

The preliminary planning also allow to formulate a more accurate budget framework. Harari had predicted early on that the cost of development would exceed the 780 million NIS mentioned in the preliminary planning stage (this preliminary amount was determined only to represent a starting point, which would allow the team to take on this large, complex mission). Now that the planners had much more data, the specification was much more accurate and allowed them to price each system and subsystem, presenting a better-founded budget estimate.

Budget constraints always stood in the background of the work. Development costs, as well as the price of producing a series of aircrafts, were taken into account. Harari: "It was clear to us that the cost had to be able to compete with the price offered by other sources. Even if our work was more expensive (the price of local production), the difference had to be reasonable. For example, if the current price of a Kfir was 7 million dollars (early eighties dollars, of course), the client would expect the price of the Lavi not to exceed 8 million dollars."

In time, it turned out that money was indeed important, but not important enough. The knowledge that the project was not limited by a fixed cost, but rather priced using the cost plus method (see above), as was common practice for defense projects during that time, allowed the project's engineers to give extra weight to technical considerations. This approach was based on the, then correct, assumption that in the end, someone would pay for the product. The team later discovered that while essentially true, that assumption had its limits: when the deviation from the original budget grew out of proportion, the project was eventually discontinued. Harari believes that had the project used fixed pricing, the IAF would have had Lavi aircrafts today.[1]

Harari: "Today, clients require life cycle cost pricing (a method that prevents surprises at the end of the road – the authors). But 30–40 years ago, we did not know what that was; everything was done en route. This approach had a defining impact on the evolution of systems engineering. In the past, defense industries emphasized performance, even if the costs were high. Today, it no longer works like that; budget planning is of fundamental importance.

When we worked on the Lavi project, we did consider the financial aspect. But we kept it in the background, and never gave it much weight, even though I knew that money was a crucial factor. This is a lesson I implemented in the very next project I managed.

Not only should the project manager and his chief systems engineer (in the Lavi Project, Harari filled both positions. For more on this issue, see above – the authors) maintain the balance between the technical and financial needs, but each and every systems engineer has to take the budget issue into account in his work. A systems engineer must have financial, as well as technological training. He must understand that money is a vital parameter. If a systems engineer focuses solely on technology, he will not have the balance required of a good systems engineer. A systems engineer must understand finances and know how to combine technology and money correctly.

Not only is this necessary when considering the aircraft as a whole, it is also a vital element in the design of each and every subsystem. There has to be a balance between technology and money. After all, the systems engineer who oversees the planning of the radar also has to work with a specified budget allocated to the development of that particular system. He has to make decisions within the limitations of the financial constraints. He must make tradeoffs between the level of technology, the budget and the schedule. He must meet the financial goals. Use of the cost-plus method decreases the responsibility of the one performing the project. During those years, the client would ask for a change, and the change would be made. Both the client's representative and the performing company knew that the costs would be covered. Such things could never happen today."

We illustrate this principle by providing an example: suppose a client wishes to include an upgraded radar system in the aircraft, without considering the cost of such an upgrade. It is then the duty of any responsible systems engineer to raise the price issue before the client. He cannot be satisfied with saying "this is what the client wants"; checking whether the upgrade is technically possible is not enough. If, after all the factors are considered, it is decided to proceed with the change, the desired activity is repriced.

[1] According to Menachem Shmul, the chief test pilot of the Lavi aircraft, who also approved the final text for this chapter (following the premature death of Professor Harari), the project was not discontinued due to financial considerations, but because the IDF preferred to receive new planes from the United States. That way, the funds allocated to the project could be assigned to other arms of the IDF.

The Detailed Planning Stage

1982 marked the beginning of the "detailed planning phase" of The Lavi Project. This was the start of the practical work, when the plans that had been devised in the preliminary planning phase were executed and implemented. The detailed planning included numerous experiments that tested the Lavi and all its subsystems. These were lengthy operations. Thus, for instance, if a subsystem failed a test, other solutions needed to be considered before retesting.

About Creativity and the Methodology: During this phase of the project, the distinction between the theoretical systems engineer and the problem solver systems engineer became clearer. Methodological systems engineering stood at the front of the stage. Creativity, which had been so important in the early stages of the project, stepped aside to make room for methodical, carefully organized work.

Harari: "In the early days of the project, during the development stages, a systems engineer has to be creative, because that is when the big mistakes are made – when there are few people and little money. This is the riskiest part of a project. Therefore, people need to be allowed to express their ideas and dedicate a considerable amount of time to the exploration and serious consideration of their suggestions. One must keep an open mind at this point, to avoid getting stuck on just one thought pattern. Later, as the project takes shape, creativity has to be suppressed, so as not to interfere with the mission-oriented approach the project needs at that point – methodicalness and order become paramount.

But even then, one must not be closed to brilliant ideas. This is not simple; it takes a lot of patience. Not all managers are capable of being patient when an employee shows up at their office after a hard day's work of searching for solutions to the various problems that come up, only to discuss something new and creative that had not been included in their work scheme. The decision of at which particular point in time the manager needs to start aiming for a methodical, rather than creative approach is dependent on that manager's character. If he suppresses creativity too soon, he might suffer great losses and end up paying a lot of money for it."

The question of creativity and the timing of its suppression raise the following dilemma: if the detailed planning stage requires a different skill set than the preliminary planning stage, should not some of the team be replaced? Perhaps preliminary planning requires more theoretical, creative systems engineers, while the more advanced stages would benefit from more practical "men of action."

Harari: "Keeping the same people throughout the project is ideal. But in an eight years long project, like the Lavi was – that is unlikely; it is only natural for some people to leave over the length of such a period. In a project that takes two-three years to complete, however, it is best if the people who started the project also finish it. The main reason for this is that when a systems engineer joins a project at an advanced stage, he does not feel committed to the actions of his predecessor. He took no part in the planning, and so it will be relatively easy for him to reject things during implementation. For instance, if a trial fails, he will tend to blame his predecessor. He will find a way to explain why the previously suggested solution was faulty. In contrast, if the one who designed the system also takes part in testing it and something does not work, that man will not be able to sleep at night. He is the one who defined the solution that would meet the client's need, he has nobody else to blame. This man will work day and night to find a solution. He is committed to the project; he identifies with it; the project is his baby."

Managing Changes

In any large project, certainly one as large as the Lavi, which lasted years, it is only natural that every now and then, demands would arise for changes. It stands to reason that sudden demands for change in plans that have taken months, if not years, to form would create quite a fuss.

Harari: "Demands for change arise all the time. When a requirement is made, one can only hope that it will not be for large scale changes. For instance, if the client demands that the flight range be changed from 300 km to 500 km, then he is essentially asking for an entirely different plane. But if he asks for a different radar capability, it is not as fundamental. It is similar to the comparison between adding a drift system to a car and replacing its 2 liter engine with a 2.5 engine with a continuously variable transmission.

With that in mind, it can certainly be said that changes are the silent killers of any project. Any change sets the project back a ways, and if the change is a big one – the setback can equal a year or more."

How does one handle change requirements raised by the client?

Harari: "In a fixed price contract, we examine the implications of the change, and present the bill to the client. The client has to know that the change he wants will cost him another two million NIS, will postpone the application phase by a month and a half, will increase the weight of the plane and decrease its flight range. Then, being aware of all these parameters, he can decide whether to proceed with the change.

A cost-plus contract on the other hand, is an entirely different matter. There, you need not worry about the budget. You are told: 'No problem, give us a bill for the change and you will get your money,' and then the change is implemented. The client is the boss, and you cannot refuse him.

But even then, a good systems engineer would tell his client: 'I want to understand the reasons for the change you want to make. Maybe then I can find a better way to meet your needs.' A good systems engineer would seek to understand the reasoning behind the client's request, and perhaps convince him that his need can be met in a better way than the one he has in mind.

The two teams, the one from the client's side and the one from the project's administration, sit down and discuss the issue together. Then the project teams study and analyze the need, formulate a response (in the professional jargon this task is referred to as QFD – Quality Function Deployment) and present it to the client. At the end of this process, a plan for the implementation of the change is devised.

This is why one of the most important qualities a systems engineer can have is learning ability. A systems engineer needs to be able to learn, to expand his mind, even at the age of 45. For example, the client raises a demand for an electric flight control system. You cannot tell him: 'no, I will revert to a mechanic flight control system, because that is what we know.' You open up the books, you take some courses, and you begin to understand radar systems."

Either way, to minimize the need for changes, one must invest a considerable amount of time in detailed planning during the project's early stages.

Harari: "One of the systems engineer's most important duties is to receive the client's requirements, carefully analyze them and make sure they are all met. This needs to be done as early on as possible, at the very beginning of the project.

An obvious principle states that the further along the project is when changes are implemented, the more costly those changes become. A large part of all project glitches occurs due to the project managers' failure to really understand the needs of their clients, and as a result, failing to provide them with a satisfactory solution.

This is exactly why good systems engineering is so important in the early stages of a project. A project stands or falls on its preliminary planning. The problem is that, even today, many people do not know enough about systems engineering and project management, and instead focus on the money and try to speed the process along. These are not just the project managers and their superiors, but the client's own representatives, who say: 'Trim it down, time's a wasting, move it along.' And on his part, the project manager finds it very difficult to postpone the performance of a certain step towards the accomplishment of his task, when it also means postponing the arrival of a ten million dollar payment."

Harari distinguishes between two types of change: "One type of change is the kind that depends on the management of a project, who failed to understand the client's requirements. There were several such incidents in the eighties, when project managers thought they had understood the requirements, only to discover later that they had not, and so were forced to turn back. The client made no new requirements; he only wanted to get what they had agreed on in the first place. For example, say he wanted the aircraft to have a lighting quality that allowed it to fly by starlight, like the Kfir; then suddenly it turned out it was not enough to meet the needs of the Lavi fighter, and changes had to be made. This was a methodological problem that no longer exists in today's systems engineering. Today's methodology clearly defines each requirement and how it should be met, demanding that the solution be presented now, rather than later.

The second type of change is the kind initiated by the client. For instance, SA-17 missiles have entered into play and the client wants the aircraft to be able to take them on. What should we do? If the project were still in its early stages, the change could be considered. But if the requirement came up during detailed planning, it would be unacceptable, because the cost would be too high."

Today's systems engineering offers a solution that minimizes the damage of such changes by creating built-in flexibility in the aircraft's systems. This approach was used in the Lavi project as well.

Harari: "We built a potential for expansion into the planes, allowing their capabilities to be upgraded. Today, this is considered common practice. Blocks of potential change are constructed, using forward thinking. For example, if the client asks for a 50% increase in the aircraft's computing capability, he does not expect to see the changes in the very first plane produced."

2.1.3 THE END OF THE PROJECT AND FURTHER INSIGHTS

The Lavi Project was never completed, but it came fairly close. In late 1986, the first Lavi prototype flew across the skies of Israel, but the question of the project's discontinuation was being debated months prior to that flight.

In the end, budget considerations (some of which came as a result of the use of the cost-plus method, which created a deep budget hole) alongside internal and external politics (the internal considerations had to do with the upcoming elections and power struggles between the Ministry of Defense, the IDF, and the Air Force, while the external ones were related to Israel's reciprocal relations with the US government) tipped the scales. IAI was preparing to terminate thousands of employees.

Harari was appointed another position, where he had to handle the management of another large project. This time around, he came prepared as a much more experienced and mature project manager and systems engineer. Lessons he had learned from his management of the Lavi project were implemented in the new project right away. Some of these lessons have been discussed here. Next are some additional insights Harari has to offer on systems engineering.

The Essence and Evolution of the Systems Engineer Profession

- A systems engineer is, first and foremost, a technical man, who has to deal with lateral, technical management issues. The professional-technical aspect has always existed in the industry (systems engineering being a field with strong ties to the industry). The new aspects systems engineering has brought with it are aspects of lateral management, because today's systems are more complex and more convoluted. Projects are also getting larger (and subsequently more expensive and more complex) and demand a more methodical, hierarchical approach.
- The work of a systems engineer can be divided into three areas: management related, professional, and technical.
  Some examples of each area include:
  1. Management related – a functional disassembly of the project: constructing the schedule, formulating the budget, and devising a configuration control plan.
  2. Professional – lateral: integrating the safety plan, maintenance plan, testing plan, risk mitigating plan, and quality control plan. Overseeing the quality of the entire aircraft.

3. Professional – technical: authoring requirements and compliance documents, designing system architecture, making engineering trade-offs, raising test requirements, and devising integrated tests.

- Developments in R&D have increased the demand for systems engineers. Development systems need the multidisciplinary connectivity that characterizes systems engineering. As a result, systems engineering has mostly evolved in industries where projects are large and complex: defense, aviation, space, software, energy (nuclear), and medical systems. Developments in systems engineering received a substantial boost in the 1990s, as a result of the accelerated development of the software field. In hi-tech, the development component is very dominant. Other industries are including more and more software in their projects as well.

The Required Skills

- A systems engineer needs to know how to handle a range of areas. For instance, in aeronautics, he needs to have knowledge of the structure of a plane, computers, and radars.
- A systems engineer is a methodological person. He has a method and the tools to use it. He must adopt process-oriented thinking. A systems engineer must finish the requirements document before he begins planning.
- A systems engineer must have a basic grasp of all the areas the project concerns itself with, because he needs to be able to ask the right questions. If I sit in a meeting at the power company, barring common sense questions, I will not know what to ask. In a meeting concerning the development of an aircraft, however, I will have all the right questions.

The Required Competences

- A systems engineer has to be a multidisciplinary individual, a man of many talents, able to talk to the experts in a clear, simple language.
- Leadership and teamwork are more important for a systems engineer than professional leadership. Soft qualities are vital; otherwise, people simply will not follow him. These qualities are needed to motivate people and handle crisis situations with them. A systems engineer needs to be a people person – a sociopath can never be a systems engineer.

- A successful systems engineer is an integrator who knows how to combine engineering with management abilities.
- A systems engineer must see the whole picture and use common sense to filter out the less important details; otherwise, the endless dive into the small details will disrupt his work processes.
- A systems engineer has to be able to quickly jump from subject to subject.

2.2

THE IRON DOME PROJECT – DEVELOPMENT UNDER FIRE

An Interview with AA

The Iron Dome is a mobile system for intercepting short-range missiles that has been in operation since 2011 and is a critical component in Israel's defense against the rockets fired at it from the Gaza strip.

Developed by Rafael, the system has gained much media exposure for operational reasons, when it showed impressive interception capabilities in an entirely new area of operation. The system had also received praise for the ingenuity of its development, seeing as it marked a new accomplishment in the minimization of development costs and was completed in an unprecedented short span of time when compared to other systems of its type. In 2012, its developers were awarded the Israel Defense Award.

One of the main reasons for this project's success was the superb management of the project in general, and of its systems engineering in particular, by three key figures: the project manager, its chief systems engineer, and the chief systems engineer of its central development array. Among other things, this array included the development of the interceptor missile. The latter of the three, AA (who wished to remain anonymous) is the interviewee in this chapter. We spoke with him about the project as a whole and about its excellent systems engineering.

Managing and Engineering Complex Technological Systems, First Edition.
Avigdor Zonnenshain and Shuki Stauber.

2.2.1 BACKGROUND AND PREPARATIONS

Historical Background

One of the things that make the Iron Dome project unique is the fact that the system was to meet a concrete operational need, namely, the need to defend against short-range missiles. This, whereas, most weapon systems are made to meet future, rather than present needs. Consequently, the pressure and sense of urgency that most weapon systems development teams have to work under have less to do with operational reasons and more to do with the relationship between the length and cost of the development process.

In 2007, this issue had come up on the agenda of a committee of the Ministry of Defense, which defined the system's requirements and considered various acquisition and development alternatives. The list of requirements concerned all three of the major factors entailed in a project of this type: cost, schedule, and performance.

The pricing requirement stemmed mostly from the need for an interceptor missile, the cost of which is relatively low. This economical consideration is especially important when the system is expected to launch a great number of such missiles against a correspondingly great number of rockets, sometimes in large barrages. The price of an interceptor is measured in proportion to the damage risked with no interception system in place, including injuries and loss of human lives.

The demand for a fast-paced development was, as aforementioned, brought on by the need to provide a solution to an existing problem. The required performance standard was beyond anything achievable by existing systems.

AA specifies: "Tests were held in order to check whether existing systems or any components thereof could be of use in the new project. One type of system considered for this purpose was aircraft interception systems. An in-depth analysis showed performance gaps in the interception accuracy parameter and in the ability to destroy the target completely. Additionally, the price of such systems by far exceeded the limits posed by the requirements."

Another example was the Spyder anti-aircraft missile system. An aircraft is a much larger target than a rocket. Its successful interception, therefore, requires a system with very different specifications. Firstly, the missile can be less accurate and still effective, even if it misses the exact target by a few meters. Second, a warhead meant to destroy an aircraft is nothing like one meant to intercept a rocket – the two have different accuracy and lethality rates.

A third air defense system considered for use was the Vulcan. Mounted on battleships, the Vulcan is a rapid fire cannon that shoots thousands of bullets per minute. The Vulcan was found to have a number of limitations as far as our needs were concerned: firstly, the system was not lethal enough; second, its interception range was too short (only 1–2 km), which meant many interceptions occurring above the area the system was meant to protect. In addition, the single system's small coverage area would require a large number of systems to be deployed in a relatively small area.

The Ministry of Defense considered other alternatives, including the acquisition of existing systems from overseas and the use of other interception technologies, such as lasers, but none of them were found to provide adequate solutions, and the Ministry chose the option of local development. Several companies applied for the job, and Rafael's offer was accepted.

AA believes the Ministry of Defense selected Rafael as the contractor, among other reasons, because its offer had two main advantages over those of the competitors. The first was that Rafael was willing to work for a fixed price (for an explanation on the fixed price strategy, see chapter on the Lavi Project) at the risk of losing money. After all, there was no guarantee that the system requirements could be fully met within the limits of the agreed upon budget. Second, Rafael had successfully convinced the Ministry that it possessed the best potential for producing a solution that would be guaranteed not only to hit the targets, but also to eliminate them completely; and this, in a very short time.

AA: "The challenges were not only in developing the interceptor missile, but also in developing the ground based system itself. This system had to be able to quickly detect the threats – which had very short life spans that ranged from several dozen to only a few seconds. Within this timeframe, the system then had to identify the target and decide what to do, namely, determine whether the target posed a threat to the defended area, and if so, calculate an interception plan and execute it by firing a missile at the target. All this, while taking into account the time it took for the interceptor to reach the target, and the need to intercept the rocket at a safe distance from its designated impact area. This set of demands created a very difficult timeline problem. These dilemmas had been raised even before the project was underway, and were, subsequently, fully resolved later on."

AA finds that the risk of losing money, which stemmed from Rafael's willingness to work for a fixed price, had come with a major advantage, which later proved instrumental in the project's success: freedom of action.

Explanation: even today, when ordering projects at a fixed price is common practice (unlike before, when the prevalent pricing strategy for the development of defense projects was cost-plus), there are still quite a few defense projects priced using a combination of both fixed-cost and cost-plus strategies. This is because in many cases, the risk level is so high that even the buyer accepts that the developing company cannot afford to take it all on itself. Naturally, in these projects, the client's representatives, fearing an uncontrolled increase in expenses, wish to maintain a certain level of involvement throughout the project. In the case of the Iron Dome, however, the bulk of the risk was borne by Rafael, and so its managers had a free hand, which allowed them to employ unorthodox methods to achieve the desired results, both in terms of the quality and number of employees recruited for the project and in terms of the quality and efficiency of the managerial and technical decision-making processes. (Note that this does not mean that the client's representatives were completely out of the picture. In fact, quite the opposite; it does, however, mean that their presence was a supporting, rather than a controlling, supervising factor. For more on this, see next.)

Preparations

The first step of preparations was to determine the structure of the project's management. It was decided that the Iron Dome would be headed by the Air-to-Air Directorate. In the early 2000s, this directorate had recognized the business potential of short- and medium-range air defense systems, and so, resources were invested in the evolving area of air defense. This investment also came in the form of making adjustments in air-to-air missiles, an area in which the directorate acquired considerable knowledge and many years of experience (the aforementioned Spyder System was the result of one such development).

The Setup and the Selection of Project Managers

The next step was determining the organizational structure of the project's advance setup framework. A preliminary distinction was made between the central development array, which included the interceptors, the launchers, and the casings (most of the development efforts were invested in the interceptor), and the remaining components of the ground-based system, which included the radar, communications, and control units. Beside the head of the program who focused on the broad management of the project stood two of the very best professionals in Rafael: H, the head of the interceptor, launcher, and casings development array; and D, the chief systems engineer for the entire program (shortly after work on the project had begun, our interviewee, AA, joined these two as the chief systems engineer of the interceptor, launcher, and casings array).

It should be noted that although the Iron Dome was developed as part of the Air-to-Air Directorate, the people who led the program came from various organizational units throughout Rafael.

During the project's early stages, it was perceived as one that did not require high technical capabilities, compared to Rafael's other flagship projects. A missile meant to counter aircrafts and ballistic rockets is nothing like a mere rocket interceptor. In other words, it was only natural that, not seeing the project as a high-end technological challenge, the engineers at the directorate were not lining up to join it.

AA demonstrates: "When people started talking to me about joining the Iron Dome, a colleague from another project met with me and said 'come to us, ours is a serious project, cutting-edge technology.' That was the attitude towards the Iron Dome. But after a while, when the development began to pick up speed, things changed, and the people in Rafael began to see the Iron Dome's significance."

This was true for Rafael's lower ranks. The executives, on the other hand, were well aware of the professional and management-related challenges, as well as the risk they had taken upon themselves. Consequently, they wanted the very best people within their reach at the time to lead the project. So, it came about that the three professional leaders of the Iron Dome program, H, D, and AA, were brought in from outside the Air-to-Air Directorate.

The reason we chose to mention this fact is that AA sees this as a major advantage that contributed to the project's success: "H came from the systems division, and he had experience in project management. Although the projects he had headed were unrelated to missile development, he did have experience in tools that included a kind of interceptor. In retrospect, he turned out to have been an excellent choice, because he saw things differently, which allowed him to be free of the restricting influence of past approaches, other projects and people. (This principle is demonstrated later in this chapter, in the example of the choice of servo.)"

H was the first to be recruited (to head the development of the interceptor, launcher, and casings, as aforesaid). He then recruited the next key position holder, the chief system engineer for the entire program, D, who had come from the missiles division where, in the years before the Iron Dome, he had mostly dealt with air-to-surface systems. AA describes D as a systems engineer of extraordinary talent.

We asked him to elaborate on what made D such a great systems engineer.

AA: "The ability to completely separate his professional agenda from his ego; although he has exceptional professional capabilities, he never becomes entrenched in prejudice. As a systems engineer, this approach allows him to have a dialogue with a wide range of people, some of them young, some more experienced, some who think like him, others who do not; and create a dynamic that leads to the right places.

Sometimes we deal with questions we have no answers for; problems, to which we see no solution from where we stand in time (unlike formal work procedures, wherein you know that if you take a certain path, you will can expect a certain result). In these situations, it is necessary to create the process that leads to a successful solution. D is able to create a dynamic that eventually leads to results – a dynamic that combines professional and intellectual skills with an egoless ability to listen."

First Steps

H and D were preparing for action. Preliminary preparations were carried out in collaboration with AK, a talented and experienced missile expert. Together, they made several key decisions, such as placing not only performance, but also costs and scheduling on top of their priority list. This meant that every engineer who worked on the project knew that when he was deriving the requirements for the various arrays in the system, cost and schedule were no less important than performance. This decision is not to be taken lightly, even in fixed price projects, because it goes against the engineers' natural tendency to ensure perfect performance first.

AA: "As engineers, if we do not meet performance requirements, we will feel we failed; if we do not keep up with the schedule, we may feel uncomfortable; but if we go over-budget, our egos won't feel a tingle. In many projects, engineers are told that out of the trinity of schedule, performance and cost they have to pick the top two. In The Iron Dome, all three were deemed equally important. Adherence to schedule and cost requirements was strictly maintained. This approach allowed the creation of a missile that was much cheaper than others in its category, and able to meet the performance objectives without going overboard with excessive requirements."

He gives an example of a "technological" key decision made by H and D, which concerned the type of servomechanism ("servo") that controlled the missile's steering system: "We began by planning the general configuration of a missile that resembled an air-to-air missile, and then asked several fundamental questions, in order for it to meet the requirements of our project. One of these questions was which 'servo' we should use for the steering system. The 'servo' must move the missile's maneuvering fins with accuracy and finesse. We deliberated whether the 'servo' should be pneumatic (a technology based on compressed gas) or electric. The Air-to-Air Directorate had always used pneumatic servos, because their properties suited air-to-air missiles. Over the years, the developers had come to believe that if a missile needed a fast, light-weight, low-volume steering system, the 'servo' had to be pneumatic. On the other hand, the Air-to Surface Directorate made bigger missiles, which traditionally used electric servos. These are usually larger and heavier, but can be easier to realize, require less design rotations, and are easier to model and incorporate into the missile's control system. They are also easier to work with in the laboratory, during the testing phase. An electric 'servo' needs only to be plugged into a computer, and testing can begin. With a pneumatic servo, testing is more complicated. It is powered by high pressurized gas, and so testing it in a laboratory requires a special support system and careful adherence to safety protocols (it is reasonable to assume that this was another reason this decision was deemed so important: time was of the essence in the Iron Dome project, and a more complex testing process would have had an impact on the schedule – the authors)."

H and D decided on an electric 'servo'. Their choice caused what AA referred to as a "sensation," because it was considered unorthodox in the Air-to-Air Administration, where the project was being developed.

Why use such an extreme term to describe what appears to be a practical, professional decision?

AA: "Because, in a way, it ignored the vast pool of knowledge Air-to-Air had accumulated over the years; knowledge, which was of great value and relevance to our project, too. Moreover, the challenges of the Iron Dome project resembled problems the Air-to-Air Directorate had handled in the past; and there they were, offering a different solution than what had always been used. Situations like these raise questions like: 'is the new solution really a good one? And if it is, how come we have never tried it before?' This certainly was an extraordinary decision, and it demonstrates the advantage of having a project manager who comes from outside the directorate, and does not feel bound by its past decisions."

However, AA stresses that the directorate heads examined the decisions of The Iron Dome managers on the merits and did not let other considerations, insofar as they arose, affect their judgment. The eventual result of this unique design process was a simple electric servo that met the cost, volume, weight, and performance requirements.

The next phase was forming a preliminary setup team. This was when AA, a relatively young electronics engineer, who had begun working for Rafael in the year 2000, was approached about joining the Iron Dome project.

AA: "At the time, I was a development team leader at the Engineering and R&D Administration. My work focused on missiles and control. D and I had met when we worked on an air-to-surface system together. When the Iron Dome directorate was beginning to take shape, I was leading an algorithm group in the Air-to-Surface Directorate and D would consult with me from time to time, as he worked on the preliminary concept. Seeing as the project I had been working on was nearing its completion, he offered that I join the setup team as an engineer. Incidentally, two days before D approached me with the offer to join the project, I had seen the rockets fired at Sderot on the weekend's newscast, said 'it cannot go on this way,' and decided that come Sunday, I would go and see how I could contribute to the war effort. That Sunday was the day I met D and got the offer to join the project as an interception engineer."

By then, a small group had already been recruited for the project. Among its members were a software engineer, an electronics engineer, a structure engineer, and an integration engineer. This was the team that created the system's preliminary design.

AA: "At that point, we mostly focused on the interceptor missile because that part of the project entailed most of the risks, and they had to be addressed as early-on as possible in order for the project to keep up with the schedule. Additionally, it was important to present the risks we had mapped as soon as possible, and show that some of them could be eliminated as early as during the laboratory testing phase, in partial system trials. At the same time, we had to derive requirements and develop arrays."

Several months later, the first test called a "Configuration Test" (CNT – Configuration and Navigation Test) was held to assess some of the project's risks. In the early stages of the project, the goal was to launch a missile that resembled the final product as much as possible, as soon as possible. This could be done because quite a few parts and arrays (such as the "servo" prototype described earlier) were already in existence and had only to be incorporated into the system with various adjustments. So, a preliminary prototype was made and launched in order to test the concept. Then, based on the test results, we could verify and remap the risks with better accuracy, identify new risks, and discover which risks did not call for as much concern as we had thought. These results created a new foundation for continued development.

After the CNT, AA became the chief systems engineer of the interceptor, launcher, and casings development array, despite having no experience as a systems engineer. He had, however, had "some experience with systems," as he puts it: "I had led a fairly large group of people. That kind a job always has some systemic aspects. What's more, the unit I worked in, the missile and control group, was a systemic group by definition, engaged in performance, algorithms, software, aerodynamics, testing, and test planning. A considerable part of Rafael's systems engineers rise from among the members of that group."

The number of people who worked on the project grew steadily, even in the supporting administration.

AA: "In the R&D and engineering administration, groups were forming, specializing in missiles and control, navigation, aerodynamics, mechanics, electronics and integration, equipping, thrust, warheads and software."

Under D's leadership, there also began a rapid development of the ground battery array, including the radar, communications, and the command and control car.

2.2.2 THE MANAGEMENT OF THE PROJECT

An analysis of AA's words leads us to the conclusion that the Iron Dome's success stemmed from, among other things, a combination of three key components: the project's organizational structure, the atmosphere among the work teams, and the work patterns. These components embodied the project's "lateral" systems engineering, a combination of technology and people management; according to AA: "Systems engineering is the central pivot that coordinates and times performance."

Lateral Systems Engineering

The nature of systems engineers is to make connections between subsystems. The more connections there are, the more complex the system and the higher the risk of malfunctions. It is, therefore, logical to aspire to minimize the number of connections in order to simplify the system (in this case, the organizational system). Accordingly, the philosophy behind the management of the Iron Dome was what AA refers to as "small systems engineering." He mostly refers to the lateral activity, which was managed by only three people throughout the entire project, even when it employed hundreds of engineers.

AA: "Compact systems engineering causes all the knowledge to be accessible all the time. We did have systems engineers for each field, like software, electronics or mechanics; but the entirety of the technological activity was managed by only two people: D and myself (this suggests that D and AA were interdisciplinary systems engineers. The other systems engineers in the project worked within the discipline they had specialized in – the authors). Our work connections ensured that there was not one detail in the whole project that we were unable to figure out either between the two of us, or by consulting one or two of the experts in the relevant field. So when a problem in a certain area needed resolving, we only needed a total of three or four people to be present at the discussion. This allowed us to find solutions more quickly."

There are more than a few projects where the equivalent team numbers 10 or more systems engineers. Why use such large teams, if they are so obviously less effective?

AA: "There are a few reasons for this. One of the more important factors this depends on is the people selected to lead the project. If the project is led by people who are able to address a wide range of issues and maintain a good, balanced relationship that allows the project's lateral activities to proceed smoothly, a small group of managers can suffice. A good choice of systems engineers in the early stages of the project is what makes small systems engineering possible."

Sometimes, the increase in the number of systems engineers working on the project happens further down the road. The project starts off with a small group of systems engineers, which grows as the project progresses.

AA: "I have seen cases where people suddenly see a need that another systems engineer could meet, and then appoint someone to run that whole discipline; then, they discover another need and find someone else and so on. At first glance, this seems like the right things to do; everyone takes care of their own area. But the problem is that in time, you lose the ability to take in the entire picture with one glance. And then you have to have a discussion with 10 other people to find the root of a problem."

The Atmosphere Among the Work Teams

One of the reasons for the project's success was, in AA's opinion, the pleasant atmosphere that prevailed in it, which created an open, egoless social environment. Very experienced people worked alongside young people who brought in fresh, different perspectives. There was no "cast system" of senior and junior employees; rather, there was a synergetic, harmonious air of practicality.

Another factor that characterized the way the Iron Dome was run was the unwavering commitment of the people to its success. There were a number of reasons for this commitment: the main reason was, of course, the sense of urgency.

AA: "The people in other Rafael projects also showed commitment to the success of their missions, and some of those projects may even have been more important for Israel, but here we had an immediate problem that needed to be resolved. In the past, I had been involved in important projects that worked on responses to future threats; but in The Iron Dome, we were working on a problem we were experiencing every day. We saw the people being fired at on television and understood the urgency. We were given a chance to solve a very serious problem in a very short time-span. This was how the Iron Dome people talked. They identified with the goal and were committed to the mission."

On this matter, he adds: "The client's demand for a tight schedule was no less important than the operational requirements. This was not a common situation. Although there are other projects where scheduling is important, the stated reasons for this are usually that 'it's a milestone' and that 'the division made a commitment before the client.' While these reasons are certainly valid, they are not enough to create the level of commitment we saw in The Iron Dome. The gruesome reality greatly contributed to the people's commitment to keep up with the schedule."

Another reason we have already mentioned in the discussion about the choice of project managers was the business risk.

AA: "The management of the company and the heads of the division (Rafael's missile division, the organizational body in charge of the Air-to-Air Directorate – the authors) understood that if they did not fulfill their obligations, there would be considerable losses. Rafael had taken a great risk, and this meant that the key positions in the project had to me manned by people who were very good at their jobs."

It follows that the quality of the people, both as professionals and as human beings, was instrumental in the project's success and the atmosphere that prevailed in it.

AA expands on the aspect of professional ability: "For a project to succeed, 'stars' need to be placed in key positions, critical junctions throughout the project. These are usually the development team leaders; they are the ones who provide solutions to complex problems. It is a matter of making the right strategic, organizational decisions, placing the best people in a project, the success of which is important. Throughout the duration of the project, when we needed someone to help us with a certain problem, Rafael assigned us the very best people at its disposal."

The Work Patterns

The systemic structure of the Iron Dome project, its management methods, and work atmosphere led to the formation of work patterns based on a principle of continuous improvement, up to the point where the result met the requirements.

AA demonstrates: "Once the price of the interceptor had been determined, a design-to-cost process began. This meant that the price became a top-priority requirement in the development of each array. This kind of process needs to be the approach from day one, because decreasing costs retroactively is nigh? impossible. We divided the interceptor into several arrays and estimated the relative cost of each one. The assessment was done in comparison to other missiles. For example, let's assume that the steering system costs 10% of the value of the missile. If the missile is considerably cheaper than other missiles, then the price of the 'servo' is extraordinarily low. Additionally, H (the head of the main development array – the authors) determined that the 'servo' should have no more than ten components. People argued that that was an impossible requirement, because it was a complex mechanism, but H persisted, saying 'for this price – no more than ten components.' Then, the engineer who led the 'servo' development group came to H's office and said: 'I have a preliminary design.' H asked him: 'Did you meet the cost requirement? How many components do you have?' the leader answered: 'twenty one components, still not sure on the price.' H said: 'come back when you get to the required price and manage to have ten components.' The leader insisted: 'But I'm already here, come see what we've done.' H: 'Not until you have ten components.' A few weeks later, the leader came back with a servo that met the price requirement and had only ten components.

Later, I got a chance to talk about this with the head of the department that developed the servomechanisms for Rafael. He told me: 'this servo is something I have wanted to make all my life. It's so simple and works so well; it is easy to assemble and also very cheap.'

We could have gotten a more complex 'servo' in one development rotation. But creating a 'servo' that simple required more thought and more talent. Finding a complicated solution is fairly easy. To find a simple solution, one had to start thinking. This example demonstrates the method used to develop other parts of the Iron Dome as well. It shows the tremendous effort invested in making each and every array as simple as it could be."

This example also shows how a lean budget can send a team in search of better solutions.

Does this mean that the products of a tightly budgeted project are superior to those of a well-funded one?

AA: "Only when the people working on it are very good; especially the leaders and members of the development groups. In my opinion, they are more important to have around than talented managers. In many cases, when a project manages to overcome a particularly difficult hurdle, the solution is not found through the use of management manipulations, but by brilliant minds; at times, it is discovered by one man who figures out how to solve the problem. Sometimes you look at the team and realize that without each and every one of those people, the project would have been stuck."

Working with the Client's Representatives

We have already mentioned that the Iron Dome being a fixed price project allowed its managers to give the developers a high degree of freedom. This was because the level of the involvement of the representatives of the client, the Ministry of Defense, was relatively low. But AA stresses that the relations between the client's representatives and the developers were very cooperative; among other things, because of the personalities of the people involved on both sides: "It did not feel like we were on the opposite sides of one process, a supplier and the client criticizing him; it felt like teamwork. They would come to see how they could fit into our teams and contribute. For instance, we needed certain intelligence information in order to make a decision, and they made sure we got it. We needed a space to perform a trial in a military base – they took care of it. They were another mind on the team, regardless of their professional background, because in brainstorming, what matters is common sense.

The Iron Dome is an example of a project where the client nearly merged with the supplier. This is not to be taken for granted, and some would even criticize it, because perhaps the client should keep his distance, so as to be able to represent the other side's point of view. If the client cannot do that, there is nobody left to give criticism. But in the case of the Iron Dome, it worked out great, because of the client's representatives, who maintained the independence of their thought processes."

AA admits that one cannot predict the results of such collaborations. But one cannot argue with the fact that the human composition of the Iron Dome made it work extremely well.

Working with the Users' Representatives

Another contributing factor to the project's success was the involvement of Air Force officials in the system's early development stages, mostly in matters that concerned its use. The Iron Dome people understood that they had to take into account not only the missile's accuracy and efficiency, but also the usability of the whole system. Thus, for instance, once the launcher and control car designs were complete, Air Force field personnel tested the system and practiced using it. They assessed its performance and suggested improvements.

AA: "The improvements made due to the operational experience of the Air Force personnel allowed us to make the final adjustments. These are critical for the system to be used comfortably in the field. Later on, the Air Force expressed its great satisfaction with how comfortable the system was to operate. In addition, this collaborative work pattern was instrumental in the quick introduction of the system to operational use."

Working with the Production Array Representatives

In many projects, the production process does not begin until the development process is complete. At first glance, this appears to be reasonable, because in the early stages of the project, developers focus on finding ways to meet the system's operational requirements. However, sometimes there are constraints that arise as a result of the limitations of the production process. In these cases, many developers say that they would not be able to divert their attention to the requirements and challenges of production until the main development risks were eliminated and leave the production specialists to their own devices. Of course, this approach lengthens the time it takes to complete the project.

When asked why, then, are the production people not included in the early development stages so that they can help prevent some of the problems production has to handle later on and save time, AA responds: "This requires resources and attention that the developers are often short on. Think of a development engineer who has a difficult problem with a certain array. He does not know how to solve it and he invests a lot of resources into finding the answer. He goes to sleep with the problem and he wakes up with the problem. On Saturday, instead of playing with his kids in the swimming pool, he thinks about whether it should be solved this way or that way. Then suddenly, someone from Production shows up and says: 'three or four years from now, when we start manufacturing the system, I'm going to want to add a few more minor requirements.' So, although taking production constraints into account sounds like the right thing to do in hindsight; in real time, it is anything but that. Additionally, in certain cases, discrepancies are formed between certain production requirements and the existing design, and additional thought is needed to settle them."

"Production and development are two very different worlds," says AA. "Each of them has its own thought patterns and its own priorities."

Not so for the Iron Dome. Here, too, the project managers chose an unorthodox approach. The pressing schedule pushed them to include people from Production early on, during the advance development stage. Moreover, the production process itself had begun before the final testing phase was over. This decision may have been risky, but the involvement of Production in the development stages minimized the risk levels.

AA shows how working together with Production had contributed to the success of the project, by helping the project meet one of its most crucial requirements – to keep up with the schedule and make the system operational as quickly as possible (some of the details in some of these examples have been purposefully changed for confidentiality reasons – the authors): "Suppose that Production deems it important to divide the missile into modular units. The developers, on the other hand, might find it more convenient to put the whole missile into one sleeve and leave the Production people to figure out how to handle it. This was not the way things were done at the Iron Dome, where the insights of Production carried a lot of weight. We discovered that with a little effort early on, during the design stages, we could achieve a tremendous advantage later, during production.

A good example of this is the time the Iron Dome's missile arrayment (the incorporation of the various arrays into one missile) had taken to complete – roughly one hour. Considering the fact that similar missiles usually take one to two days to array, this did not only save the project considerable time, but quite a bit of money."

The Iron Dome's managers were fortunate, because the production specialists they had cooperated with were, in AA's words: "Excellent production experts and engineers, who brought unique perspectives to the project and helped us find solutions."

He demonstrates: "We designed an optical component that had parts glued to its circumference in a very specific pattern. The developers devised an 8 hour long assembly process. The process was so lengthy because once one part was glued on, it was necessary to wait until the adhesive was partly dry before proceeding to the next part. An experienced production technician said that our process took too long and suggested another way. His process took less than an hour. It was wonderful, especially for a system like the Iron Dome, where processes constantly needed to be simplified to save time and costs. Had the situation been different, they might have told us 'why are you wasting your effort on this? The developers have thought of everything and their process is correct. Please follow procedure."

AA summarizes: "Rafael had learned that the inclusion of Production in the early stages is a reasonable price to pay for the benefits it yields. Today, they are trying to implement this lesson in other projects."

Checks and Balances – Making Trade-Offs

One of the most obvious expressions of a systems engineer's work is the creation of checks and balances between interconnected systems and subsystems. In the professional jargon, these are called trade-offs. AA agrees with this statement (which appears in many of the chapters of this book) wholeheartedly. He holds that it is a systems engineer's most important mission and gives an example that illustrates how this work requires both professional knowledge and a broad, systemic perspective. These, in AA's view, are the things that make a good systems engineer: "In the early design stages, we allocate requirements to the different arrays. The assumption is that when the arrays are later put together, the requirements will be met. We also often include a safety margin, so that a deviation in a single array does not bring down the whole concept. The problems start when one of the arrays cannot perform up to par. When that happens, one option is to persist – to continue investing efforts and funds until the array meets the requirements. Driven by their egos and the desire to overcome the challenge, this is what engineers naturally tend to do.

This is where systems engineering comes in. It examines how another array can make up for the deviation; especially if that array has been completed with no hitches, meets the requirements, and can deliver even more than it currently does. If a certain array presents a difficulty, we first check whether a reasonable effort has been invested to make it meet the requirements, or whether a grievous error had taken place. Here too, it is good to ask ourselves whether we have arrived at the 'knee-point,' the point where even a slight improvement requires a lot of effort, which, of course means it is advisable to go no further. The next step is to consider how we can make up for the difficulty by improving other arrays – this is an acceptable solution in many situations."

He demonstrates by describing a specific case (some of the details of which have been changed for confidentiality reasons – the authors), when a systemic process of balancing different disciplines produced a satisfactory solution ("satisfactory, but not optimal," he stresses) for the operational system: "The Iron Dome system has a sharp, cone shaped component called a 'discard ogive,' part of the interceptor missile. When the interceptor approaches the target, the 'ogive' is discarded and the homing head is exposed. This part of the system entails a very complex development process – not only does this process have to take place while the missile is flying and even maneuvering through the air at great speed, the discarding of the 'ogive' brings about a sudden change in the missile's aerodynamic configuration and subsequently changes its behavior. Moreover, the ogive's separation and distancing from the missile must be perfectly smooth; it cannot bump against the missile, not even enough to leave a scratch. And to make matters even more complicated, the discard must be possible under various flight conditions: at low or high speeds and at various air densities (which change depending on altitude and the maneuvers the missile is performing). Of course, the 'ogive' also has to be very cheap and sturdy, and has to possess certain aerodynamic properties. These were the requirements, and the engineers had to rack their brains for the answer.

Now, when you look at the process, you see the conflicts of interests: what works in one situation does not work in another. When we began to develop the array, we soon discovered that the mission was a 'challenge,' which really is mild word for 'impossible.'

So what did we do? First of all, we mapped the problem, noting the things we could improve upon easily, and the ones we would have to struggle with. Next, we had to reexamine the requirements, in case some of them would turn out to be irrelevant. For instance, the missile would only be able to make sharp maneuvers at low altitudes, because at higher altitudes, there would not be enough air for sharp maneuvers. There was, therefore, no need for the missile to be able to perform sharp maneuvers and withstand low aerodynamic pressure at the same time. This narrowed our range of requirements down to the relevant situations and sets of requirements. If the team is still unable to meet the requirements at this point, it moves on to the next step.

Another example: we had a conflict of interests between the strength of the 'ogive,' which we achieved by adding layers and making it thicker, and the aerodynamics – which were negatively impacted by the ogive's increased sturdiness. Now we had to decide how thick we could make the 'ogive,' so that it could withstand both the extreme conditions of its discard and the impact to its aerodynamic structure, these being only 2–3% of the cases.

We made a tradeoff between strength, aerodynamics, and mechanics (the discarding mechanism); and to that we added the algorithms that had to make up for the gap we had been unable to bridge (so that the missile could still withstand extreme conditions). And so, we arrived at the 'knee-point' – the point where any further improvement would bring only negligible benefits. From there on out, it was no longer worthwhile to invest in this area."

On Models, Simulations and Testing

The more complex the system, the less one can safely rely on human intuition to make it work. Complex systems need to be periodically examined from the earliest stages of their development, in order to locate limitations and failings that advanced planning had failed to foresee. This is why the system simulation exists – to test the design, consider alternatives, plan tests, research performance and check various algorithms.

AA: "The quality of the simulation is measured by two main criteria: how close the simulation is to real world physics, and whether the simulation includes most of the subsystems that make up the main system. Making a good simulation is an art-form. Is entails the creation of a flexible, modular structure, able to handle numerous components that change throughout the development process. The structure of a simulation must allow it to test a system that grows continuously, as more and more new components are added, and as existing components are enhanced."

This suggests that the construction of a simulation also requires trade-offs: "You have to decide which phenomena or components to model, and how accurate and detailed that model should be. If the model is very detailed, you can simulate more phenomena and get closer to reality; but high levels of detail cost precious time and money. So it is important to use common sense when deciding what level of detail should be considered adequate. A simulation only has to include the most influential, most important phenomena. The development of the simulation is a continuous process that relies on learning from tests or field activities."

Evidently, the Iron Dome had set a high bar here, as well. The project used systemic models and simulations extensively. The prevalent opinion in Rafael is that this element was vital to the successful development of the system in such a short time and its rapid entry into operational use.

A similar approach was adopted in the planning and operation of the testing array: "The development process included many tests. We tried to hold a test every few months. We tested many different scenarios, some of which even lay outside the system's performance range. Our goal was to locate weaknesses and limitations and make improvements and corrections accordingly. Large numbers of tests are a contributing factor to success, but they also make the system more expensive. However, the low price of the interceptors and of the Iron Dome targets allowed for a relatively large number of tests."

To conclude, the success of the Iron Dome project – an efficient technological system developed from scratch in a remarkably short time, while remaining within the limits of a strict budget, was achieved thanks to a human conglomerate of project managers and engineers; a conglomerate that had the ingenuity to use creative systems engineering and form a challenging and highly committed work environment. Another factor that contributed to the project's success was the fact that its developers were allowed to test the system's function in "real life," seeing as the IDF had used it even before it was fully approved for operational use.

PART III

THE INTERVIEWS

3.1

DEVELOPMENTS IN A COMPLEX, TECHNOLOGICAL WORLD – THE AVIATION AND SPACE INDUSTRIES

3.1.1 "STRUCTURED, MULTIDISCIPLINARY METHODS OF RESOLVING LATERAL PROBLEMS"

An Interview with Norman Augustine

The government of the United States engages in diverse and widespread activities at the South Pole. According to the American perception, the southern continent of Antarctica should not be under the jurisdiction of any particular country; it belongs to the world. True to this position, the United States established a research station at the South Pole, making no territorial claims and has been cooperating with various countries in conducting research studies of the continent. These studies range from investigations into the origin of the universe to the impact of global warming on sea-level rise.

In addition to the station's research objectives, its location also bears geo-political significance: the government of the United States bases its viewpoint on the assumption that if they were to leave such a strategic location, other countries might seek control over the area – possibly even leading to military conflict. A number of nations have in fact made claims of ownership of (sometimes overlapping) parts of Antarctica, nearly all of which include the South Pole area. Therefore, in order to advance science and prevent such a situation with its sensitive implications from occurring,

Managing and Engineering Complex Technological Systems, First Edition.
Avigdor Zonnenshain and Shuki Stauber.

the government of the United States has been operating and financing the research station and its support stations for decades. These stations located in Antarctica are provided logistics from locations beyond the boundaries of the continent.

The US activity was assessed during the second half of the 1990s, particularly due to the fact that operational costs were extremely high – a natural result of long supply lines and severe weather conditions. The examination was conducted by a committee of renowned experts who represented a variety of relevant disciplines. Mr. Norman Augustine, an expert in Systems Engineering and Aeronautical Engineering, headed the committee, and at the same time, served as president and CEO of the large aerospace manufacturer, Lockheed Martin. The committee published its conclusions in 1997, in principle: to continue the research operations and activities and to find ways to improve efficiency and reduce costs. This resulted in the construction of a new station at the South Pole to replace the then-existing station that was deemed no longer safe for habitation.

More than a decade passed, and in 2011, another committee was appointed to examine ways for reducing operational costs. Norman Augustine headed this committee as well, even though he had already retired and had been devoting his professional skills to diverse public service activities.

The conversation with Augustine focused on two primary subjects: the working and behavioral patterns implemented by the committee while attempting to make the United States government's operations in Antarctica more efficient – working patterns drawn from the content world of systems engineering; and his perception regarding the essence of systems engineering.

Examination of Operations in Antarctica

Augustine believes that he was not chosen to head the examination process because of his experience as head of the committee in 1997, which had conducted a broader examination and had reviewed implications and policies. In contrast, the current committee was being summoned to examine performance: the task was to examine ways for reducing logistical costs of the research studies being conducted in Antarctica.

Augustine: "This task is more related to my skills as a systems engineer – my experience as someone who has managed wide-ranging projects, which necessitate the formation of groups from different fields of activity and coordinating work efforts.

The following is an example of a dilemma that most people would not consider, whereas a systems engineer probably would. What would happen if the length of the shower that each member of the research station team at the South Pole took could be shortened by 30 seconds? Why would this be significant? Because there is plenty of ice there, but not liquid water. The ice, of course, must be heated and melted into water – a process that consumes large amounts of energy. And it is not only the cost of the fuel used for heating, but the distance it must travel in order to reach the station that is also of significance. Fuel is brought to the station by ship that requires the service of icebreakers and then by cargo aircraft. People at the station understandably might not give much thought to the 30 seconds. They simply want to take showers.

We wanted to evaluate the fully burdened cost of a shower per minute. This is an example of a systems engineering analysis. In addition to the technical, engineering side, there are also human aspects. How will people react to the idea of taking shorter showers? If they are told that the reasons are simply for saving on expenses, they might object. But, if they are told that the money saved will enable them to conduct additional research, they might be motivated, as scientists, to agree. Psychology also enters the picture, as does economics, and the wider perspective of systems engineering is required. But the problem of linking showers to icebreakers to aircraft is further complicated in that some energy must be consumed to generate electric power to provide light and heat for the station – and if that process produces excess ("waste") energy that can be used to help melt ice, then the equation changes altogether."

Forming the Staff. The committee operated within the framework of the "National Science Foundation and the White House office of Technology Policy" that also provided the members of the committee with a supporting, administrative team. The first step in selecting suitable committee members was the preparation of a list of some 25 subjects that, in Augustine's estimation and in accordance with the opinions of White House and Foundation personnel, were the central issues to be addressed by the committee. For example: how to employ icebreakers in order to reach the South Pole; how to provide for proper health-care services; how to prevent fire in an environment with lots of water – but all in the form of ice. (This past year a Brazilian station was destroyed by fire.)

The next stage involved the characterization of the skills required in order to cope with the issues being raised. For example, the ability to conduct a financial analysis, an understanding of ice breaker operations, and the input of someone who had endured weather conditions of 80° below 0 °F.

Augustine: "We reached the stage of identifying those with the necessary qualifications. We noted the names of people we were acquainted with, as well as the names of those most likely to know people endowed with the skills that we were looking for. We also made use of the Internet. We contacted 'The National Academy' and reviewed the names of scientists and engineers. We inquired into who excels in this field and who in that one, who was willing to devote the considerable effort needed, including travel to the South Pole. We received recommendation for the most qualified individuals and contacted them to see if they might be willing to take part in the mission." A matrix of project needs/individual capabilities was the result.

The search and detection team did not review the candidates merely by their expertise and professional reputation: "Personal qualities were also important and the qualities of the recommended candidates were examined. We did not want to appoint individuals who might be 'zealots,' with decisive and firm positions that were not subject to change even in the face of new information. From our point of view, this was crucial.

The second important quality that we looked for was the ability to work with other people. This kind of framework calls for numerous meetings and discussions and lots of hours together on ships, on airplanes and in conference rooms. When you have one individual who wants to do all the talking, the system breaks down. We were looking for people who would want to listen at least as much as they would want to talk. On the other hand, you don't want a room full of people who just sit there and don't say a word. You want those who know how to be contributors."

Members of the committee were expected to do their jobs voluntarily as a public service. (The committee was unpaid.) In Augustine's estimation, it would be a matter of 8–9 months of part-time work. The willingness among the members to participate in the committee's activities stemmed from a variety of reasons: the desire to take part in an undertaking of national importance (most committee members were involved in numerous such activities;) out of interest and the challenges involved; the desire to be part of a unique experience; and the desire to make a contribution.

Why wasn't an undertaking of this type imposed upon an international consulting firm like Mackenzie?

Augustine: "A consulting firm would not have been able to put together a group of professionals of the caliber of our group. For example, if Mackenzie had contacted me, I don't think I would have accepted. I am not looking for traditional, paid work at this stage in my life. But, when the folks from the White House asked – I willingly accepted,[1] considering doing so part of one's duty as a citizen."

Planning. After the committee members were chosen, a working framework was formulate. Augustine: "I made an outline of the final report and we began allocating responsibilities to each member of the committee. By and large, those were executive decisions, since we needed to balance the use of the particular talents of the 12 of us. As the work process progressed, we made a number of revisions when we found that some 'chapters' overlapped or that new issues arose. We could generally deal with this by combining task groups or adding new ones.

As we got further into our undertaking we realized that we had focused so much of our attention on activities in Antarctica itself and its first-tier supply points (New Zealand and Chile) that we were not adequately examining the "base" activities back in the United States. New task groups were created to specifically focus on the latter. These were composed of members designated by the chairman based on their individual skill sets."

Gathering Information. The next stage was the gathering of information in order to characterize and decompose the Antarctic support system and the ways in which it functioned. This included visits to the various research stations and their supporting

[1]The resulting committee of 12 included the recently retired head of the US Coast Guard; a recently retired "four-star" Air Force General; the recently retired chairman of the US House of Representatives Committee on Science and Energy; the recently retired head of logistics for one of the world's largest international corporations; a retired Navy Admiral; and a number of highly respected scientists and engineers.

bases as well as meetings with the people involved in the research activities at the South Pole and other sites – some housing over 1000 people in small villages and others consisting of a tent or two.

Since members of the committee reside at different locations throughout the United States, as well as a representative of the French Antarctic Program, we periodically got together for 2-day meetings (there have been six such meetings to date), not including our extended travel time together. We generally met in the evening of day one and continued discussions throughout the following 2 days. During these meetings, experts and professionals with relevant information appeared before the committee. For example: an expert who addressed us was able to shed light on the function of traverse vehicles under extremely cold climate conditions.

The committee considered both specialization matters as well as more complex and comprehensive systemic subjects. One of the more important issues was the matter of shipping, handling, and the transportation of goods.

Augustine: "Large amounts of supplies reach the South Pole by ski-equipped, four-engine turboprop aircraft. Special vehicles (that move on caterpillar tracks) can also be used to reach the South Pole, but it takes quite a long time, 15 days for a round trip from the coast. This is a complex choice with many different aspects. It is the analysis of a system, which is the focus of systems engineering. In order to decide which manner of transportation to choose, both the advantages and the disadvantages of each option need to be evaluated, as well as unintended consequences of making changes, thus involving everything from weather analysis to human safety and the impact of operations on the natural environment. Therefore, all of the committee members take part in these discussions. On the other hand, the matter of which type of aircraft is more appropriate is a specialization issue that does not require the input of the icebreaker expert. For the same reason, not everyone goes to visit all of the different sites. Sometimes, only a group of three to four committee members makes the trip to a particular location. The committee members were, as a whole, already quite familiar with the issues at hand having collectively made 82 trips to Antarctica, of which 16 were to the South Pole."

What Is the Connection Between the Project and Systems Engineering?

One could certainly ask the question: Why should a description of a project of this kind be included in a book that deals with systems engineering? It appears to be a classic example of organizational consultation, including the gathering and analysis of data, the pinpointing of problems and recommendations for improvement. It is not the design for a new airplane.

Augustine: "It is the characterization and analysis of a very large and complex operation comprised of numerous components that interact with one another and require a process of trades-offs, similar to the example mentioned earlier regarding methods of transportation. Another example: perhaps it would be wiser to integrate robotic sensors and a broadband communications system instead of flying some of the researchers to Antarctica at all. This way, some of the scientists could remain at home and fewer people would be needed for the on-site collection of data which is a major driver of cost. In fact, 90% of the person-days spent in Antarctica as part of the US effort there are involved in logistic support of the 10% who are actually performing the research."

In other words, the thought processes of a systems engineer were required in order to achieve appropriate results. According to Augustine, even though most of committee members do not consider themselves to be systems engineers, the vast majority of them are endowed with the ability to broaden their thought processes and think by the rules of systems engineering in dealing with problematic and diverse situations.

About Systems Engineering

What Is Systems Engineering? A system is anything evolved from elements that need to work together and that affect one another. Systems engineering is the art and science of assembling numerous components together (including people) in order to perform useful functions. For example: an air logistics system includes air traffic control radars, airplanes, pilots, passengers, runways, communications, meteorologists, and much more.

- It is not merely the assembly of components, but rather the assembly of components in a coordinated fashion. The concept is not to assemble components and then await the outcome; rather, it is to assemble components in a logical, efficient and economical way that accomplishes a desired function (without interfering with the performance of the functions of other systems).
- Systems engineering is a way of dealing with wide-ranging problems in an organized and disciplined manner.
- Engineers generally solve engineering problems, as opposed to systems engineers who deal with problems that involve many diverse factors ranging from people to economics and from politics to science.
- Systems engineers deal with problems by synthesis, analysis, and trade-offs.
- One of the most important areas of activity for a systems engineer concerns trade-offs: determining the appropriate "dose" of each component in the system.
- Systems engineering is a managerial tool. But, it is more than engineering management – it also deals with matters that are not "physical." In many aspects, it is not a technical skill. This can be frustrating for engineers who do not like dealing with human issues and their ambiguities.

Not Just Engineers

- Systems engineering thus differs from traditional engineering in that it also takes non-engineering components into consideration. For example: economics. Economics is a central component in systems engineering.

Systems engineering has grown from within engineering because both fields of study are based on similar principles, such as the assembly of components, their analysis, trade-off, and modeling. These tools are also useful for solving problems in areas beyond the fields of hardware and software, such as in industrial organization design, health-care management, and financial investing.

- The word engineering has become a generic term related to problem solving. We are familiar with terms like organizational engineering.
- Engineers generally know how to deal with the laws of nature: if you do things exactly the same way in the same environment, then whatever works today will usually work tomorrow. On the other hand, if people are included in the system, the result might be very different. The identification of systems engineering with the world of engineering does not necessarily stem from the fact that engineers are required to undertake much of the work, rather from the fact that the discipline has grown from within the field of engineering. For example: the input of economists and financial analysts has been most significant in many of the projects in which I have been involved in my career. Today, the combination of engineers and physicians is having an enormous impact on health care.
- While at the Pentagon in 1965, a Colonel with whom I worked conducted analyses of the likely contribution to national security of funds assigned to ballistic missiles as compared with submarines, civil defense, airplanes, or training. He was criticized for this "audacity," but, in essence, we make exactly these trade-offs every year when we prepare the defense budget. But we do it in our heads, often without disciplined methodology or analyses. Systems engineering can provide discipline in building a budget: how much should be allocated to health, to national security, and to roadways?
- Human Resources Managers often apply systems engineering tools, but for traditional reasons they are not defined as systems engineers; system engineering is usually classified among the physical fields. The thought processes of systems engineering are taught in depth in engineering courses, but not in most other fields of study.
- To some, systems engineering is not considered a specific discipline, rather simply as a means for dealing with complex problems.

Skills of a Systems Engineer

- In order to be a good systems engineer, one needs first to specialize in a single professional field and then broaden oneself to become a systems engineer. This is because systems engineering touches upon problems that are extraordinarily complex. Solving these problems requires a variety of skills – more than one person alone is capable of developing. But almost all good systems engineers previously had established themselves working in a single discipline.
- A good systems engineer needs to combine understanding of fundamental phenomena, experience with a wide range of fields, with a deep and comprehensive understanding of one professional field. Many different people have

worked for me in my career: lawyers, financial personnel, chemists, engineers, people in marketing and advertising, as well as those involved in human resources. Knowing how to put these pieces together is simply an example of systems engineering in a different context. For example, at Lockheed Martin, there were 180,000 of us working together to accomplish challenging goals.

Leadership and Systems Engineering. Many believe that a good systems engineer often needs to display leadership qualities, particularly when in charge of goal-oriented teams that are working with a limited budget and a tight schedule. Norman Augustine supports this belief. He writes and gives lectures about leadership quite frequently, while emphasizing the elements that touch upon the work of systems engineers:

- Leaders cause the right things to be done. They seek out the good, the just and the correct result. Therefore, Hitler and Lenin were not leaders, even though they motivated – or intimated – people to act. That is not sufficient. Some of the greatest criminals in the world had leadership qualities, but they lacked integrity – true leaders motivate people to achieve inspirational and constructive goals and do so because they want to do so.
- While conducting my research study on great leaders, including those that I have been fortunate enough to know, I discovered that in spite of the numerous differences in their individual personalities and styles, they share a number of similar qualities. These qualities were revealed particularly during times of crisis. As the saying goes, during times of calm, every ship has a good captain. I found 12 common traits among the finest leaders I knew:
  a. *Character* – if people do not trust you, they will not follow you. True leaders have high moral and ethical standards. They do the proper thing, no matter what the price.
  b. *Vision* – leaders know what they are aspiring to achieve ... in the words of GE's Jack Welch, they can see around corners.
  c. *Competence* – a person cannot lead the way without being knowledgeable in his or her field of activity. Among the many important skills required for leadership is the ability to evaluate people and be able to decide who is more suitable in a certain situation. This is related to my philosophy of management: find good people, tell them what you want and step aside.
  d. *Energy*: leaders are people committed to their goals. They never run out of energy.
  e. *Courage* – a leader needs to instill confidence in others. He or she must have the ability and the daring to take calculated risks.
  f. *Perseverance* – a leader needs to display determination, resolve and persistence. Never give up ... unless it becomes clear that the goal being sought is no longer appropriate.

g. *The ability to motivate others* – a leader motivates people to do things above and beyond what they thought they were capable of doing. In this world of ours, which is very achievement-oriented, the ability to make people invest that additional effort is what makes the difference. A leader motivates by setting an example. In addition, a leader has to show people that he or she believes in them, and enable them to express their creativity. This is an essential condition for building a team.

h. *Lacking egotism* – Great leaders are team players. They give credit to others. This is how they achieve great accomplishments: by being goal-oriented rather than being motivated by personal interests. Ironically, the best way to get ahead is not to try to get ahead.

i. *Decisiveness* – a leader needs to be capable of making the tough decisions ... sometimes without all the facts they would like.

j. *Judgment* – decisiveness needs to be combined with good judgment. Leadership includes intuition and a kind of control mechanism that prevents one from making impulsive decisions. This quality is reinforced by the ability to take criticism and to listen to others and to learn. According to the dictionary, judgment is "a process of formulating an opinion or an assessment by means of diagnosis or comparison." Good judgment is the ability to recognize the best of possible approaches to a challenge ... and the ability to let go of an outdated one.

k. *Mentoring* – great leaders create other great leaders. One of the more outstanding ways to characterize leaders is to examine the people that develop under their mentorship. Leaders give encouragement, inspiration, and advice.

l. *Listening* – Peter Drucker said that 60% of all managerial-administrative failures stem from a lack of communication. A leader must know how to listen. Warren Buffet told me that leaders need someone next them to remind them that "the Emperor has no clothes."

3.1.2 "PLANNING SYSTEMS THAT FIT THE NEEDS OF BOTH CLIENTS AND USERS"

An Interview with Yossi Ackerman

A question that still has not been answered clearly is whether systems engineering is, first and foremost, the concern of engineers, who see it as a tool for handling complex systems; or a systemic management field that affects management patterns and non-technological organizations as well.

This chapter explores the positions of the president of a major technological corporation on this issue. An engineer by training, Yossi Ackerman has never filled the position of a systems engineer, but nonetheless sees himself as one – an engineer who thinks systemic.

Background

When most of his friends would go to study at the local high school of his hometown in the Galilee, Yossi Ackerman would get up two hours earlier to get to Haifa, to Bosmat Electronics Tech School. His parents had decided that the boy had to learn a trade and become an electronics technician. Back then, in the 1960s, Bosmat was a reputable practical and theoretical school. The Technion, for instance, would receive its graduates with no entrance examinations.

Yossi Ackerman: "We studied advanced theoretical electronics and mathematics. Alongside that, we worked in workshops and learned mechanical and microchip processing. The school's teaching philosophy was multidisciplinary. We learned to use a lathe to produce accurately shaped and sized pieces; we learned how to document our work, so that we could continue from the same point the next day; we learned about mechanical drawings, but we had bible studies too – because an electronics engineer has to know more than just electronics."

Upon graduating, Yossi Ackerman was accepted into the IDF's Academic Reserve program and studied Aeronautical Engineering at the Technion. He describes it as a systemic discipline, an aggregate of mechanical engineering, control, and other fields.

During his military service in a flight testing unit in the Air Force, Yossi Ackerman's career in systems did not halt its progress: "The good thing about that unit was that we learned new things all the time. They would tell me 'we want to receive a new electronic warfare system,' and I had to learn all about it, approve it for use by the air force, train the team of users and integrate it into the air force. At that point, I knew nothing about the area and had to start studying it from scratch. Later, there would be a new bearing that they wanted to use on the plane. So I went from one field to the next, and discovered that everything I had learned in Bosmat and thought to be irrelevant was in fact very relevant, because they had, in effect, taught me not to fear anything new. So today, whenever I stand before a new subject, I know I can handle it, or, rather, that I can learn to handle it. And if, after that, there is still something I do not know, I bring in an expert who does."

After being discharged from the military, he returned home, to manage the family farmstead. For 7 years, his main occupation was raising turkeys. Once again, he had to learn new things, and once again he employed the approach he had adapted for himself – what he did not know, he either learned or used the services of experts: "I did everything myself. I was a welder; I built turkey coops, installed air conditioning systems, engaged in biology, food engineering, economics, and the management of expenses and income."

Yossi Ackerman had obviously left an impression on the friends he had made during his high school, university and military years, because in 1982, after 7 years, in which he had not practiced engineering, he was approached by Elbit Systems and recruited – not as a junior engineer but as the head of the Lavi Project (for a more detailed discussion on the Lavi Project, see interview with Ovadia Harari), in which Elbit was in charge of the plane's avionics (e.g., the navigation and piloting systems, or the weapon arming systems). After a period of hesitation, he accepted and, at first, started working for the company in a half-time position.

The first month was, unsurprisingly, traumatic, and, towards the end of it, Ackerman asked to be released from the mission. He describes the conversation that followed with his superior rather graphically, seeing as it had been a constituting event in his professional life: "After three weeks, I came to my boss, Yossi Tidhar, and said: 'We both made fatal errors, you in recruiting me, and me in accepting. Let's erase everything; I'm going back to the farm. The people here are much smarter than me, they know everything, and I don't understand what they want from me.'

He said to me: 'Let's sit down. Tell me what your difficulties are and we will resolve them. Let's take care of each problem separately.'

I said: 'First of all, I don't understand all these letters: SOW, WBS, MFD.'

He replied: 'When you are sitting with other people and you hear something you do not understand, write it down on a piece of paper, then later come to me and I will explain what it is. There aren't thousands of these words, maybe thirty. What is the next difficulty?'

I said: 'In three days, I have negotiations in Tel Aviv with the CEO of a certain company Elbit has been in conflict with for a long time, and I do not even know what the argument is about, not to mention Elbit's position on the issue. How can I be sent there?'

And he answered: 'Let's break this one down, too. Who do you think can handle the problem?

I said: 'This person knows the economic issue.'

He said: 'Excellent, he will join you at the meeting. Who else knows anything?'

I said: 'That person knows the technical part, and the one who headed the project before me knows the history of the conflict.'

He said: 'Both of them will join you, then. Prepare for the meeting with the three of them and go together, If anything else that you are not familiar with comes up at the meeting, tell them you need time to think and will get back to them later with an answer.'

And that it exactly what happened. The negotiations went smoothly, and the end result was excellent. I have carried this lesson with me ever since. This is how I resolve crises. Whenever there is a difficulty, we break it down into components and address them one by one."

Yossi Ackerman had been the manager of the Lavi Project until it was shut down in 1985. That year, he replaced his superior as the manager in charge of the aviation projects in the company's air division (which included production systems and other elements as well). As time passed, he moved up through the ranks, until, in 1996, he was promoted to the position of Company President, which he filled until the year 2013.

He sees himself as a systems engineer, although that had never been his job title: "I think like a systems engineer. Even as a farmer, I was a systems engineer. Back then, I did not know I was practicing systems engineering, because the term did not exist. Anyone with a can-do attitude is a systems engineer. This is systems thinking, and if you add the word 'engineer,' you just narrow it down to the technological field. All managers' work is essentially systemic."

Systems Engineering at Elbit

As a large technological company, Elbit does everything that falls within its areas of activity. Therefore, Yossi Ackerman says: "We encouraged people to handle areas outside their specializations and lead projects Elbit had never before attempted. For example: we heard that the Air Force wanted to privatize its flight school, or the Firefighters Squadron, or the UAV project (unmanned aerial vehicle – the authors). At first, we did not even know what those were. But someone took it upon themselves to handle it. They would study the subject, gather materials and read the literature. This is how we won the UAV tender, and the Firefighters Squadron Operation tender. Compared to us, a niche company with no systems engineers would say: 'This, we do not do,' because they would not have the person who would say: 'I'm a systems engineer, and I know how to learn new subjects.' In that company, the systems engineer would be the CEO."

He brings the example of a niche company Elbit had acquired: "Tadiran Kesher made only basic radios that connected five or six modules – a small, focused niche. The company had maybe one systems engineer. This was not a disadvantage – one can make a lot of money focusing on a specific field."

He adds: "This approach is relevant to any area. For example, the more specialized a doctor is, the more he charges, and the less he knows. He must also say that he knows less. He even has to admit he has no understanding in other areas, so that he can focus his expertise. On the other hand, a general practitioner, who has to provide answers to any problem, is a systems engineer. He either knows the answer, or says: 'I will read up on this and get back to you with an answer,' or sends for an expert. A general practitioner is a profession, much the same way a systems engineer is."

All along the road, we find a significant overlap between the job of a manager and that of a systems engineer. It is, therefore, not surprising that in many companies, including Elbit, there are such job titles as "Technical Manager." The boundaries between the various position holders are unclear, and Yossi Ackerman sees this as a good thing. He argues that there is no benefit in setting clear boundaries, if constraints that arise force the boundaries into place anyway: "In Elbit, those assigned to fill the position of systems engineers do not necessarily have degrees in that discipline, but they do possess the relevant skills to the project. There are projects with dozens of systems engineers, but not all of them bear that title. It is, however, likely that as we rise through the system hierarchies, there is a higher chance that those who bear the title 'systems engineer' will also have the words 'systems engineer' written in their formal job description.

There is need for flexibility. For example, if Elbit employs one hundred systems engineers, but at a certain point in time, only has enough systems engineering work for sixty of them, the remaining engineers will perform other tasks, like those of technical or marketing managers.

The same principle holds for other jobs. For instance, when the head of a division has to choose a manager for an electronics project, he can simply choose the best electronics engineer available. But, if the project includes a range of disciplines, he might want to choose someone who is not as good at electronics, but is the best at systems thinking. If that engineer is good at electronics, but unable to make the necessary compromises when faced with various constrains, then he is not suited for project management."

Yossi Ackerman finds that whether or not someone is a systems engineer throughout their professional life, even if they do not bear the title, is a question of character. To demonstrate, he brings the example of Professor Ovadia Harari (see interview with him in this book), who, even as the Vice President of IAI, remained a systems engineer: "Of course, as vice president, he was a super-systems engineer, because in a position like this, one can only delve so far into the specifics, professionally.

Defining a systems engineer is fuzzy logic. It is not something that can be defined. Nor, in fact should it. It needs to be given space, and then defined in accordance with the given situation. A manager who looks at the whole technical and technological picture can be called a systems engineer."

He agrees with the statement that the higher the rank of the engineer, the more management elements his job entails.

More on systems engineering in Elbit

On the effects of the matrix structure: "The Lavi Project had employed hundreds of people from various disciplines. But, because of Elbit's matrix system (see also the interview with Mimi Timnat), those who worked with me were not subjected to me, but allocated to me, so that at any point, their superiors could decide to pull them away into another project. A systems engineer needs to know how to make people want to work for him. In addition to the technical part, he must have an understanding of psychology – of people. Some knowledge of economics and law is recommended as well."

On the training of systems engineers: "In the 90s, hundreds of people in Elbit were already practicing systems engineering. Not all of them were called 'systems engineers,' but in time, they, too, received the title. Then, we saw that their knowledge of physics or electronics was not enough; that they had to be taught systems engineering. And so, we began to devise additional training programs for them.

A good systems engineer can manage without furthering his studies. This is true for every field: there are exceptional teachers who have never studied pedagogy, and there are, of course, those who have. There are excellent systems engineers who never attended formal training frameworks. They possess the right qualities and are self-taught.

Nonetheless, continuing education programs have an added value. They have put things in order. They poured meaning into what systems engineers were doing. These programs also create a common denominator among Elbit's systems engineers, each of whom had arrived from a different unit. As time passed, because of the growing importance of the systems engineer's role in each project, we wanted to institutionalize the field and asked the Technion to create better-founded educational frameworks that awarded Master's degrees (see interview with Professor Aviv Rosen)."

On a systems engineer's professional background: "A systems engineer needs to have a technical background. He does not have to be an engineer; he can be a technician or a practical engineer. He can also be someone who has accumulated technical experience as a technical officer at the air force, because he understands the client's needs. But these situations are rare. In most cases, a systems engineer in Elbit has basic engineering training."

The Evolution of Systems Engineering

Yossi Ackerman first began to recognize the importance of systems engineering after the cancellation of the Lavi Project, in the 1980s: "The Lavi was invented by engineers. They did not ask the pilots what would suit them best. They said: 'We know which engine is best.' The client was silent, and that is what brought this project down. So we realized we had to become customer-oriented, rather than engineering-oriented. We had to give the client what he needed. We also understood that providing the client with a worthy solution required more than one discipline."

He describes the background of the emergence of systems engineering:

"In the past, there were washing machines that required two technicians to operate. Gradually, people understood that in order for everyone to be able to use a product, human engineering was needed too. If the machine were made by an engineer alone, its design would not require systems engineering. He would create an excellent motor, but the machine would rust, break down every 2 days, and cost four times as much. To make a good washing machine, one needs to know about water, to prevent it from being covered with scale after a month of use; one needs motors; stainless steel, to keep it from rusting; one needs knowledge of electricity, economics, marketing.

Over the years, companies began to understand that more than one discipline is needed to advance a project. Kodak, for example, is among the companies that had failed to grasp this. After all, a carpenter, who failed to realize that he had to learn a thing or two about metal, lost his livelihood. Today, architects study psychology, because one needs to know more than one discipline."

On the development of the patterns of working with clients: "For many years, our American clients would give us blueprints, and we would execute them – this was called the 'Build to Print' method. Later, the method changed to 'Build to Spec' – we were given a specification and asked to make a card that had all the functions listed in the specification. This method was also used widely for quite a few years. I introduced a different method, called 'Build to Need.' It means, understanding what the client needs, not what he wants. Some clients think they are very clever, and they will tell you exactly what is needed. In these cases, we do not have much left to do, and we lose our creativity. We would receive a specification: a thick book, a thousand pages long. We did not need to prove that our product was good, just that it met the specification requirements.

As time passed, clients understood that they needed to make their books thinner, so they explain what is needed, not how to do it. They can tell us that if a rocket is launched at us from the Gaza Strip, they want our system to intercept it within ten seconds, at a given distance from its target. This is fine, but they should not tell us whether the launching vehicle has to have eight wheels, or six wheels. Telling us that the operating unit has to number eighteen soldiers is also legitimate, but they should not plan everything."

On the direction systems engineering is headed: "One of the main problems is the unwillingness of experts to talk to others from outside their discipline. A doctor who specializes in internal medicine will not consult an orthopedist. If a patient hospitalized in the internal medicine ward has an orthopedic problem, he will be taken to the orthopedics ward, because the orthopedist will not come to the internal medicine ward. If, god forbid, the patient dies because of an internal problem, then, as far as the orthopedist is concerned, he died healthy, because 'his part' was fine. This is the very same reason economists failed to predict the economic crises, and political scientists failed to predict the Arab Spring. Had they all thought together, in interdisciplinary teams that included economists, political scientists and psychologists, they would have managed to predict these events.

But all this is about to change. Thanks to the internet, now everyone understands, and reads, and knows everything. Nobody trusts the experts anymore. Some of the writers in the world's leading journals today are architects, psychologists, not just scientists. It is beginning to sink in."

More Insights on Systems Engineering

On the relationship between systems engineer and project manager

- "The difference between a program manager (project manager in Elbit-speak) and a systems engineer is insubstantial. A systems engineer sees the entire project from a technical-operational standpoint. The program manager also sees the technical-operational aspect, as well as other things, such as the economic and legal aspects. For all that, there is considerable of overlap between them. A systems engineer does many things the program manager does. In small projects, one person fills both positions.
  The relations between the program manager and the systems engineer vary, according to the nature of each project and the people working on it. If the program manager is very economically inclined, the systems engineer will enter the field of economics as well, because he knows that if he creates a system that cannot be afforded, it will not be a good system. There is no structured framework defining the activity areas of each one."

On measuring a systems engineer's success

- "A systems engineer is measured by the success of his project. Uzia Galil (founder of Elron – the authors) once said: 'A great engineer is one who plans something useful that someone is willing to pay for.' If a systems engineer has made something no one needs, he is not a very good systems engineer. A good systems engineer must also understand the constraints. He needs to know that if he makes a technologically impressive system that requires 1,500 soldiers to operate it, that is not a good system."

3.1.3 "SEEING BEYOND TECHNOLOGY – UNDERSTANDING THE MISSION"

An Interview with John Thomas

Like the majority of professionals we interviewed, John Thomas, INCOSE President during 2012–2013, had already adopted the thought patterns of a systems engineer from the onset of his career as a young engineer. Like the others, he had no inkling at the time that he had done so. During the 1970s, he worked as an electronic engineer with the development teams of the US Air Force and there, he adopted the work patterns of a systems engineer – naturally – as someone who sees things not only from a technological perspective, but from a wider, systematic one.

After his work with the Air Force, John Thomas became a member of the Department of Defense industrial complex. During the 1990s, after completing his position as Head Systems Engineer within four separate programs, he joined the ranks of the large, international consulting firm, "Booz Allen Hamilton." Thomas climbed the company ladder and was promoted to serve as a Senior Vice President and Head Systems Engineer. He left the company in 2012 and founded his own consulting firm.

We spoke with him about the development of his career as a systems engineer and his viewpoints on the consolidation and nature of the discipline.

Personal Background

John Thomas is a clear example of an engineer who was not instructed or trained in the discipline of systems engineering. Rather, the systematic approach was an inherent and integral part of his basic thought processes. This approach assisted him to excel from the very start of his professional career.

He began his career as an electronic technician with the American Air Force. Until that time, the start of the 1970s, the electronic systems in these airplanes were mostly analog electronics. Technicians knew how to open damaged devices and repair the damage, or, if needed, to replace the damaged parts. This trend began to change with the development in computer capacity. The complex digital devices that had been designed were much more difficult to repair and, could not be repaired by the technician, but had to be shipped to intermediate depot locations for repair. The approach calling for the replacement of digital devices, opposed to repairing them, began to take hold as standard operating procedures for those technicians working on the aircraft.

John Thomas: "This change led to frustration among many technicians who saw their role being transformed right in front of their very eyes. They had been trained to examine the source of the electronic problem. Now they had to know how to decipher computer codes – which is very different from rendering an electro-mechanical solution."

Example:

Within the Air Force Wing in which John Thomas was employed, a problem was discovered in the inertial navigation subsystem of the FB-111 aircraft. When asked to deal with the defective inertial navigation subsystem, he did not limit his examination to a technological angle of that subsystem alone. He studied the specifications of the equipment and the ways in which the navigation subsystem had been integrated to interact with other subsystems within the aircraft. Thomas discovered that a constantly recurring problem of the navigation subsystem stemmed from the flow of signals between the navigation subsystem and other avionics subsystems related to it. He then studied the specifications of the equipment in these other avionics subsystems (e.g., the avionic flight instrumentation subsystem, and found the root cause of the problem. Flight compass signals were at times out of tolerance due to aircraft placement on the runway. These out-of-tolerance compass signals generated a failure during the start-up sequence of the inertial navigation subsystem). Thus, he was able to successfully tend to the problem, where others had failed to do so.

John Thomas: "It was not the professional knowledge, in and of itself, rather, it was my mindset. My mindset provided a means for system-level thinking that made the difference in fixing this problem. I made use of analytical tools and displayed curiosity regarding the interactions among the subsystems of the aircraft. I tried to understand how the algorithm of the larger system addressed the specific subsystem that I was working on. Even though I was an electronic technician, I did not limit my examination to the specifications of the electronic circuits. I also reviewed the sequence of activities that lead to the implementation of the task and the ways in which the different parts are integrated into it."

Later on, John Thomas studied electronic engineering and returned to work with the Air Force – this time as an engineer and an officer: "I continued working in the same manner. It was my job to solve systematic problems. If a system didn't work properly, it would be taken apart and its components would be examined in order to determine what was wrong. The algorithm would be examined and then it would be reassembled."

When did you realize that in essence you are working as a systems engineer?

John Thomas: "I did not know what a systems engineer was until the mid 1980's. We were working on certain project when members of the Department of Defense came to visit my unit in the Air Force. They presented the field on which they were working and it was quite evident that I understood the mission and technology as well as they did, and from a systems perspective, even better than they did. They recruited me to serve as the Head Systems Engineer of the project. That's how I formally changed from being someone who thinks systematically into a head systems engineer."

Systematic Thinking and the Human Perspective

John Thomas agrees with the distinction between a "pure" engineer and a systems engineer, claiming that an engineer sees the parts, whereas a systems engineer sees not only the parts, but also the connections between them that create a process. This process generates a new and valuable behavior that no one part by itself could provide. In his opinion, this type of perception is not solely a gift of nature – it can also be learned.

To clarify this position, he presents the psychological model developed by the research duo, Isabel Myers and Katherine Briggs.[2] Their highly acclaimed model, exposed in the 1970s, divides personality types into four categories of courses of action. These four categories of action are mixed and matched to form 16 combinations of personality profiles. These profiles reflect different behavioral patterns. Individual behavior can be characterized according to one of these 16 profiles. But, Myers and Briggs claim that this does not mean that a person will behave solely in accordance with the behavioral patterns that make up his personality profile. Rather, the personality profile reflects the mode of operation an individual feels most comfortable operating within. Individuals can and do operate with different personality profiles when required to do so for the success of their career. Operating outside their "preferred" personality profile can cause them stress until they master and become comfortable with the expanded set of skills required to be successful – but they can be taught these skills. Thomas feels strongly that this claim by Myers and Briggs is indeed true. He has gone through multiple training programs to make him more comfortable in operating outside his preferred personality profile. The result has been an expanded set of both soft and hard skills that have benefited Thomas's career.

In relation to systems engineering: Even though not everyone is born with the characteristics of a successful systems engineer, a person can train himself to be one. But, he might be a bit uncomfortable playing the role: "There are some people with systematic perception who notice the connections and sequences, but do not display any interest or enjoyment in doing so."

(Why would a person want to serve in a function that involves activities in which he feels "less comfortable"? Apparently, for a variety of reasons, for example: one could assume that a successful engineer, who values the importance of being promoted along the administrative hierarchy, might want to serve as a systems engineer, during an intermediary stage, in order to prove his capabilities in fulfilling this role on his way to being appointed to an executive position – the authors).

Thomas, who joined the ranks of the consulting firm "Booz Allen Hamilton" in 1991, said that if a systems engineer wants to be influential, "he has to be able to pinpoint the central track that influences the process."

He gives the following example: "We were consulting a client who was having fundamental problems with his internet protocol. We noted that the personnel spoke only in technological terms, about volume and velocity, whenever referring to the problem. I did not limit my attention to the technological problems, but also observed the body language of the client's representatives, who came from technological backgrounds. I noticed how they talked with their colleagues about the problems, both confirmed and unconfirmed. There was clearly much more at hand than mere technological problems. It might be very easy to adhere to the claim that the entire problem is technological, but that is never really the case. I asked questions and

[2] The four categories of the Myers and Briggs model: (i) The way a person processes the energy around him (sequences of introversion–extroversion); (ii) The way a person thinks (intuitive–analytical thinking); (iii) The way a person gathers data; (iv) The way a person makes decisions.
Each profile includes the four categories, in varying amounts, thus creating 16 different profiles (4×4).

noted that the responses were very dramatic. Also, everyone seemed to be addressing the problem with different terminology. It seemed as if they were talking about the same thing, but using different words. One of them called a certain phenomenon 'rust' while another called it 'brown'. Why were they behaving this way? It was just confusing matters. The explanation: each one viewed things from a completely different angle based on the different histories of their careers."

John Thomas sees himself as a systematic thinker: "A systems engineer needs the basic skills of critical thinking as well as engineering terminology. Today I see the complexity of the system more than the solution to the technical problem. Today, contrary to the past, I cannot offer a quality evaluation of an electronic panel. However, I am highly capable of technical designs. I have the basic knowledge to ask the right questions. I can tell if the person who responds to them knows what he is talking about. I have the ability *to see beyond the technology and to understand the problem.* I also have an understanding of the mission or business challenges that the technological system has to solve."

He emphasizes the importance of understanding the human component: "A good systems engineer sees not only the technical parts, but notes the human components, as well. As a systems engineer you acquire the skills to synthesize the data, to examine the mutual relations between them and to understand the considerations of the individuals within the group.

This ability enables a systems engineer to speak in terms of 'brown' with one group and in terms of 'rust' with the other. In this case, the systems engineer becomes a highly intensified interpreter. In the previous case, people were speaking a different language and they were frustrated. They came from different units within the company and were at odds with each other. We were the ones that clarified to them that they were actually talking about the same problems and were all concerned about the same things."

This incident illustrates one of the primary challenges that systems engineers are faced with – to be able to recognize the real problem in need of a solution, while in most cases it is not the problem declared by the client.

Thomas cites another example: "About ten years ago I worked with a group of people who defined themselves as systems engineers. We had been asked by the management of the organization to strengthen the abilities of their systems engineers. When we met with the group, it became apparent that the problem was not the lack of abilities among the engineers, rather the lack of understanding and agreement at the managerial level regarding the implementation of the tools available to systems engineers for solving problems. There had been a council of systems engineering, but it was abolished because senior management did not understand its relevance.

As consultants, we were asked to improve the skills of their systems engineers. In reality, the level of understanding among members of the senior management and the ways for incorporating the current staff of systems engineers were the matters in need of improvement. Once this happened, they were able to use the systems engineering base that they already had, which was actually quite good, to continue to develop. But, the next part of the assignment was not so simple. After the real problem had been identified, the group best suited to solve the problem had to be selected. The existing staff of system engineers, on whom the improvements were to be focused, was comprised of frustrated personnel who had felt unappreciated for quite some time. The problem was that they were not the right people to teach the members of the senior management how to make use of systems engineering, evidenced by the fact that they had not been able to do so during their years of employment there.

This group felt the lack of appreciation and suffered from demoralization. Therefore, the central task was to change the attitude of the group. We had to show them that their attempts to influence the management had not been effective because they had relied on only one set of tools in order to do so.

The members of the group had to realize that since they had not been successful in achieving their goals by using their current tools, it would be senseless to continue along the same pathway, doing the same things over and over again. They had to approach the problem in a different fashion. We sold them on approaching the problem in a different fashion by appealing to their greatest skill. As system engineers their great skill was to think systematically about a problem. They soon saw that they had a socio-technical problem. Not simply a technical problem. Once their viewpoint of the problem was changed by this expanded view of the system they quickly embraced different approaches to attack the problem from which they were accustomed."

If the problem is behavioral and not technical, isn't that a classic example of organizational consultation?

John Thomas: "Yes, but for a group of systems engineers. Telling them to go and check to see what the demands are, which is a clever way of saying: 'go and check to see what the real problem is,' is an approach promoted by systems engineering. It is not a technical process. In my understanding, it is the curiosity of the individual to view interconnectedness as part of the task. All along the way, the systems engineer comes into contact with human beings, in various aspects. The job of the systems engineer in these situations is to ask simple questions and to listen. This enables him to raise additional points and to examine the interconnections.

In many instances, organizational consultants are equipped with the skills of asking and listening. But, not many organizational consultants have the skills of a systems engineer of interpreting the problem from a systemic view point. In many cases, organizational consultants view these kinds of problems primarily as being problems of interpersonal dynamics."

According to John Thomas, a great amount of additional work is needed for technological organizations to understand that technological systems cannot be managed separately from human systems. The two are intertwined.

He recalls: "We discussed this during a visit to one of our clients. They said to me: 'you hit the point. We have many good engineers and many good systems engineers. However, some of our great engineers do not have the social skills required to form coalitions and work across multidisciplinary (technical and nontechnical) teams. This creates monumental problems.' We had been developing their technological abilities, but had never implemented the concept that they need to serve as facilitators by means of integrating human talent.

Wherever I go throughout the United States, and when I meet with a company vice president or a senior vice president, I find that if the discussions begin with talks about technical problems, they end up with talks about systems engineering – but in a different way – as if by hinting at the subject. Interpersonal dynamics are not even mentioned. Only the technical problems are voiced. You won't hear a word about the fact that they don't know how to work together. When asked what their fundamental problems are, they often tell you: 'My engineers are struggling to resolve a specific technical problem'– you won't hear – 'My engineers create social conflict. They don't know how to navigate through conflict to arrive at a more effective solution.'"

About Leadership and Systems Engineering

The understanding of the centrality of the human factor in the world of systems engineers led to placing an emphasis on the importance of leadership – the leadership qualities of people – as an important component of the discipline. This issue is also addressed by others interviewed for our book (see the interview with Norman Augustine, for example). John Thomas is an enthusiastic supporter of this position. He was quite alarmed when we asked him if he, in his position as Head Systems Engineer, is actually a manager.

John Thomas: "No, certainly not. I am not a manager. I am a leader. I allocate managerial responsibilities to others. Managers see people as production units. 'How many hours did you put in?' 'I did a performance evaluation on you.' In contrast, leaders see people as human beings. Leadership comes as a result of a world-view. It is not 'you did fine here' and 'you didn't meet timetables there.' A man is not a machine. Once I was that kind of manager. I had left the wounded behind and was on the verge of being fired. It was a crisis that brought about a fundamental change in me.

A leader is able to have people follow him because they want to be part of what he is leading toward. The team gives him power over them – through his authenticity and transparency of purpose. A leader sets the standards and the manner of behavior he expects to be conducted and then demonstrates by living to those standards himself. By doing so, he provides others with the fundamentals of systems engineering."

Leadership is extremely important for systems engineers because the variety of skills and actions required for developing and building a complex system is so great that only a group of individuals working together can do so. Therefore, leadership is required in order to ensure that their actions and efforts are successfully coordinated and integrated. In his lecture, John Thomas states that "technological leadership is not an option for a systems engineer, it is an absolute necessity."

The Rational and Intelligent Use of a Variety of Qualities

It can be deduced from Thomas's words that a systems engineer should be endowed with a variety of qualities – qualities that sometimes contradict each other. As such, he needs to maneuver between them. Increased awareness enables him to do so, even if the role is less comfortable for him (according to the aforementioned Myers and Briggs model).

We illustrate our meaning by listing and describing a set of rational and intelligent qualities that a good systems engineer should be endowed with, according to Thomas:

In his opinion, a systems engineer should be able *to move from domain to domain* by means of a wider perspective of the overall system. The desire to do so stems from natural curiosity and the ability to recognize situations in which it is possible to do so. In order to get into the thick of things, a systems engineer has to know how *to ask the right questions* and to delve deeper into the matter (a quality that sometimes requires a display of assertiveness and the ability to insist on receiving a fitting response). On the other hand, a systems engineer also needs to adopt a humble and modest approach, certainly during his entry into a new field, and to refrain from behaving as if he is Mr. Know-It-All. *He should acknowledge the fact that he does not know enough and, at times, admit that there are things he does not understand – and then ask for additional clarification.* (Assertive people are not fond of finding themselves in such situations and are reluctant to admit their inferiority – even it is temporary).

Isn't this similar to the situation in which the manager of an organization in a certain industry is transferred to manage an organization in another industry? Isn't it like those who claim that management is management, under any conditions? So, similarly, systems engineering is systems engineering, regardless of the situation.

John Thomas: "Whoever thinks like that is making a big mistake. Even in management it is not the same. A manager who does not understand the field in which he operates, the tasks and the products, and who thinks that management is primarily about making financial decisions, will lead the organization to disaster. If the Head Financial Manager becomes the General Manager of the company and thinks that it is simply a matter of profit and loss, he will destroy the company. On the other hand, if he invests time and energy into processes and into the development of new lines of company products, he could have a very powerful influence on the company.

You can't move from domain to domain if you lack the humility to understand that the basis of your managerial knowledge is insufficient. It will suffice only after you have studied the field in which you are to operate. A systems engineer knows how to begin to build the box, but he cannot make decisions within that box until he has collected data and learned from others. In these situations, he has to surround himself with people who will help him to absorb the basics of knowledge required to turn him into an expert. A systems engineer has professional knowledge, but he must remain humble and be aware of what he does not know. This way, he can learn a great deal and he can learn quickly.

I have been involved in multiple domains during the course of my career. I have been successful in them because, when moving from one domain to another, it took at least a year to reach the previous level of problem solving. At the start of my involvement in each domain I lacked an overall perspective of the phenomenology of the task."

Another important quality of a systems engineer is the *ability to simplify complex matters* (see the interview with Kobi Reiner):

John Thomas: "The ability to simplify things enables a systems engineer to serve as a problem solver. I am able to get to the root of the problem within a short period of time, even if I am not proficient in the field. And, I always have an expert in the field nearby. It is a combination of methodology and intuition. If you rely solely on methodology, you will not find a solution. You need a process to help you direct your intuitions."

In his lecture, John Thomas defines the six basic skills of a systems engineer as follows:

1. Craftsman: knowledgeable in the processes, work methods, and technical work tools.
2. Functional: combines the ability to delve deeply into matters with a wider perspective.
3. Programmatic Understanding: this includes the abilities of contract planning and communications, planning expenses, and the ability to assess timetables.
4. Leader: the ability to function under pressure; to manage conflicts; to make decisions; to know how to communicate, form, and lead work teams; and to be a man of vision.
5. Problem Solver: the abilities of critical thinking, systematic thinking, and associative thinking, and the ability to simplify matters.
6. The ability to understand one's surroundings: familiar with the technology and the fields, building a suitable life circle, and understanding the tasks at hand.

The Training of a Systems Engineer

Academic training programs for systems engineers are usually at the Master's Degree level of study. The widespread opinion is that only an engineer with experience in an industrial field can really learn and internalize the fundamentals of the discipline. One of the problems that the field is faced with is that basic engineering studies focus on the technological plain. We asked John Thomas, who agrees that this indeed is the situation, while he feels that engineering schools follow this line of focus.

He responded: "When learning about circuits during my studies of electronic engineering, I learned how to design an oscillator. But, I did not learn how the oscillator takes part in a broader system or how it should be situated onto an electronic panel. I did not learn how a number of such panels fulfill a much broader function. No one ever explained to me how to integrate my professional skills into a broader context. The same can be said today. When I talk about this with the Deans of Engineering Departments, their response to me is that they do not have enough time to integrate all of the engineering fields that should be taught into their academic programs. I tell them that I am not asking them to add more courses, rather to instruct the lecturers to present the context of how course work fits into a system. So that their engineers, be they chemical, mechanical or mathematical engineers, who have a natural tendency for the subject, will say: 'great, now I know how to do this technically, but I also understand how it fits into the broader picture.'

I tell them that there is no need for an additional courses on system engineering or on leadership at the under graduate level. However, I am concerned about the fact that they come into contact with so many engineers and never mention that powerful design (mechanical, chemical, electrical, ...) comes from a system perspective and that the power to influence decisions comes through leadership skill. They don't display enough systematic thought in the classroom. The important point is whether someone who finishes school is curious about the significance of his work."

3.1.4 "SIMPLIFICATION CAPABILITIES IN A COMPLEX ENVIRONMENT"

An Interview with Dr. Kobi Reiner

We have already described Rafael's activities as the developer of the Surface-to-Air Missile Defense System – "Iron Dome." This chapter will discuss the systems engineering characteristics as being expressed in one of the earlier, "air-to-surface" missile projects, developed during the late 1990s and early 2000s, while placing an emphasis on a systems engineer's professional development within the project.

Before the Project

Risk Reduction. Rafael has two sources for its knowledge creating processes. The first is "top down," achieved by financing the business directorates that perceive future operational needs. The second is "bottom up," it stems from the engineering teams, where the engineers try to improve operational response in two ways: by presenting new technical and/or algorithmic capabilities, which are then revealed in professional magazines and conferences; and by optimizing the efficiency of the processes that improve the product's performance and increase its reliability. The funding the "bottom up" activities receive is negligible, seeing as most of the initiatives that originate from the lower levels stem from the engineers' need to challenge their own intellect in an attempt to bridge their professional knowledge gaps, and the need to make development processes (which are not funded by the projects) more efficient and reliable.

In the professional jargon, these are known as "risk reduction" processes. They are called that because, when an order comes in for the development of a relevant project, where the knowledge created at the "lower levels" can be made use of, it helps minimize the technological and process-related risks that otherwise might hamper its success.

Development initiatives that originate at the lower levels occur partly due to encouragement from the organization itself, and partly due to the engineers' natural professional curiosity. For instance, an engineer can return from a professional conference, having discovered gaps in his knowledge. He feels he must now fill these gaps and go in search of solutions. Or, another engineer can find a certain technological capability possessed by competitors while browsing through professional publications, and try to duplicate it for Rafael. These initiatives are part of any engineer's professional development, and pose substantial professional and operational challenges. In many cases, such initiatives form collaborations within Rafael, or evolve within the framework of existing professional workgroups.

"Bottom up" development initiatives that receive no funding from the business administration units are known throughout Rafael as "Black R&D". Kobi Reiner, a chief systems engineer in the R&D and Engineering Directorate at the Missiles and Network Centric Warfare Division, says that a substantial part of the early "Black R&D" initiatives do indeed find their uses.

He explains and illustrates: "The general risk factor is comprised of a quantification of three risk components: a technical component, failure to adhere to the schedule, and the possibility of the risk being realized. For example, the United States has very accurate navigation systems, and we want to achieve this level of accuracy as well. During the project's launch phase, there are, as yet, no systems in Israel with that capability, but work is being done to improve the performance of existing systems. Since it is impossible to know for certain that the mission will be a success, this constitutes a risk assessment – an assessment of the company's ability to achieve the desired performance level.

The risk level is, of course, derived from the complexity of the task. If, for example, somewhere in the world, a certain new technology emerges in the field of missile guidance systems that use image matching, the task of integrating this capability into our missiles is considered relatively simple, thus making it a 'low risk' project. If, however, we want to develop an entire missile that includes the new navigation system, a new engine, new wings, a new homing head and other new elements, and then integrate all these as well, the task will be considered a 'high risk' project."

Learning of Capabilities and Creating the Technical Specification. IAF operations had defined the special operational needs for an air-to-surface missile, and presented them before those responsible for developing the armaments for IAF. This group initiated preliminary talks with representatives of the Rafael Air-to-Surface Directorate – the professional unit tasked with the development of the missile. The purpose of these talks was to assess Rafael's ability to meet the operational requirements. IAF representatives met with Rafael experts as needed (some were "risk reducers," whose developments were deemed relevant for the project) to learn about the options for each of the missile's subsystems. The output of these meetings was the definition of the missile's technical specification.

Kobi Reiner: "Risk reduction does not mean risk elimination. In this project, before the FSD (Full Scale Development – performed in accordance with the previously formulated specification), we spent a year and a half working to reduce avionic risks (the missile's avionic capabilities, such as accuracy and the ability to successfully home on the target) and aero-mechanical risks (the missile's engineering capabilities, such as the ability to perform and structurally withstand maneuvers). Nonetheless, it is important to stress that designated risk-reduction activities allow the project to enter the FSD phase with reduced, if not fully eliminated, risk factors."

The technical specification document accounts for the findings of the risk reduction phase in two ways: the capabilities demonstrated at the risk reduction phase are integrated into the document as requirements, while high-risk capabilities (for which there is no certainty of accomplishment) are omitted from it, or labeled "optional."

These pre-launch phases entail no commitment on the part of the potential buyer. However, there are cases when the Air Force chooses to invest in the project financially, even in this early stage of assessing Rafael's existing capabilities.

Kobi Reiner: "At this point, before the project is underway, the systems engineer needs to be a professional as well as an entrepreneur, intimately familiar with the professional environment, because his main goal is to define technological ability, and assess whether the project can be realized at all. This is why, in the early, risk reduction stages, it is not necessary to dote on the small details of the realization. Rather, it is important to concentrate on proving the company's technological realization capabilities.

This process can begin with the creative, slightly disorganized individual. The further into the project one gets, the more one requires a management-oriented mind. The nature of the ideal systems engineer changes as the project advances. The ability to put the right systems engineer in the right place makes the difference between success and failure."

The Project

Phase One: A Disciplinary Systems Engineer

The Organizational Structure of the Project. The new missile's development included two interwoven projects: the avionics project, and the aero-mechanics project. Both project chiefs reported to the Head of the Program. Each project included disciplinary systems engineers, each responsible for the development of all the elements in the subsystems in his area.

Kobi Reiner: "Today, the components of a system are becoming more complex and more convoluted on all the technical levels; so, understanding them requires familiarity with more engineering elements. Consequently, the systems engineer's work becomes more complicated. The multitude of engineering elements makes integration difficult."

Part of Kobi Reiner's job as a disciplinary systems engineer on the aeronautics project was managing three "leaders" at Raphael, and heading an external subcontractor's designated project. Each of the leaders served as a development engineer and was in charge of developing a component in a subsystem, for which Kobi was responsible. The "leaders" are called thus, because they are fully responsible for leading the development of the component under their charge. Each leader, in turn, manages professional development personnel: the developers.

Kobi Reiner: "The 'leaders' are not systems engineers by title, but in order to realize the component under their charge, they, too, must, in effect, do systems engineering and project management work. The integration of two or more areas is, in itself, a task for a systems engineer. For this reason, in small projects, one person can fill the position of both leader and systems engineer."

To summarize, the organizational structure of the new missile's development had five levels: the head of the program, the heads of the two projects, the disciplinary systems engineers, the leaders, and, finally, the developers.

One of the complexities of a large project is the question of organizational reporting. The further down the organizational structure one looks, the higher the chances that the employees he finds will not be an integral part of the project; rather, they will simultaneously serve the needs of other projects, and have other tasks, related to their expertise. In this case, the three most senior ranks, including the disciplinary systems engineers, were exclusive to the project. The leaders and developers, however, acted in service of the project, from within their respective professional departments.

Kobi Reiner: "I set the technical objectives for the leaders, but those who managed them were their superiors within the engineering array. In our case, in the early stages of the development phase, which demanded engineering-oriented focus from the team, all of the leaders' time was dedicated to the project, but some of the developers beneath them were assigned to work on other projects as well. A situation like that might create management constraints, caused by the developers not being sufficiently available. The leaders were not always able to resolve these problems, and so, when they believed I would be better suited to deal with the superior of the developer in question, they asked me to intervene."

On the disciplinary engineer's work methods:

Kobi Reiner: "During the initial processes of formulating the work patterns for the FSD phase, the systems engineer must be very well-organized. This is both so people can have a clear, well-defined framework of the work ahead of them, and so that the engineer himself can write the system requirements specification along with them. This activity took the form of joint work meetings. I support the team meetings approach, because it minimizes the risk of things falling between the cracks. I formulated the overarching requirements specification for the system under my charge. The three leaders then derived the requirements for the components under their charge from this specification. For this, a leader would derive the requirements for his component from the overarching specification, and then construct yet another requirements specification, which would, in turn, serve the developers working under him."

Although a disciplinary systems engineer directs a sizeable group of people, the bulk of his work is systems engineering related, mostly because said people are not directly subjected to him.

Kobi Reiner: "Most of my work was engineering and coordinating between the different persons that performed the work for me at different (physical) locations.

One person's output was another's input – for one leader to be able to move forward with his tasks, another had to complete his own task and provide the first leader with its results. My job, as a systems engineer, was to coordinate and synchronize these actions.

I was also required to coordinate laterally. Thus, for instance, from time to time, I had to check my own systems, for which I needed testing equipment, which was the responsibility of another engineer who worked on the project. He did not always find it convenient to provide me with the equipment whenever I wanted it. This and other dependency situations between the inputs and outputs of the engineers in the project are examples of the constraints I was faced with. Of course, any problem can be brought before the project manager, so that he can use his authority to resolve it. But a good systems engineer who owns his actions wants to solve his own problems. He will only turn to the project manager for these kinds of solutions as a last resort. His first duty is to try and overcome the hurdles himself."

Phase Two: Leading the Process of Obtaining the Avionic Certifications

Taking Responsibility. Kobi Reiner's advancement within the project hierarchy was not planned. Rather, it was a direct result of his work and the initiatives he had led. When a task requires a lot of coordination, as is typical of systems engineering, many things fall between the cracks. Those who bother to "pick them up" stand out for the better.

The head of the program asked Kobi Reiner to take up another, relatively small task, that did not justify assigning a full-time disciplinary systems engineer. The task was to supervise and coordinate the work of a subcontractor – IAI. In this assignment, he no longer represented his own, avionics division, but the entire development program.

The project kept moving forward, and entered the phase of the first system trial. The trial was comprised of rendering digital simulations of the future missile's performance. This required synchronization between the testing equipment, computers, and simulators.

Explanation: the approval processes of a system trial include the important stage of going over the hybrid simulation (a simulation that combines operational hardware and software – the authors). In order for the tests to be realized in the hybrid environment, there needed to be full synchronization between the electronic software specialists who had constructed the special equipment used for the hybrid simulation, the leaders of the various operational units, who had to bring the products of their work into the hybrid approval environment, the software specialists, and the algorithm specialists.

Kobi Reiner: "At this stage of creating the integration, I was more dominant than the other disciplinary systems engineers. The nature of integration is such, that you try to put things together, and they refuse to work. Not everybody has the ability to manage the big picture and deal with the small things at the same time. Some are better at planning and writing, others are better at making the connections. A systems engineer can be creative, he can have a talent for designing the architecture, but unable to lead the development of the final product. I took the responsibility because I wanted the system to succeed. I got into it naturally, without being officially asked to. I found myself in meetings with many leaders who were not assigned to me, and I managed the convergence process. I became a sort of lead systems engineer for the avionics project. Actually, I managed the technological integration."

Did that not disturb the other disciplinary engineers?

Kobi Reiner: "It was done in good spirit, because there were people there, who did not place their egos before everything else. They recognized that they were unable to do what was needed to complete the approval processes, and I thought they were glad that someone else came in and prevented their systems from failing. Of course, had someone objected, I would not have gone near his people – if he did not want help, he would get no help. But that did not happen, because the head of the program had been wise enough to select people who were both skilled professionals and amiable individuals."

In time, Kobi Reiner's actions caused him to be formally responsible for leading the avionic approvals. Here, too, he based some of his work on joint meetings, such as he had in the project's early stages: "When people talk among themselves, they fill in the gaps. Had I spoken to each of them separately, I would have risked leaving some of those gaps unfilled. On the other hand, in crowded meetings, there is always the chance of what we call 'junk requirements' coming up – someone has a less-than-relevant idea, and he believes it's important that we look into it. Then, he needs to be told, politely, that his demand is off-topic.

A systems engineer has to be able to cut, to know when he is spreading himself too thin. Engineers tend to complicate things, and it is his job to stop them, because today, the possibilities are endless. One of the key traits a successful systems engineer must possess is the ability to simplify, when the atmosphere is one of complexity and complication. A good systems engineer prevents complication from emerging."

Phase Three: Chief Systems Engineer

Three years after its launch, the structure of the program was reorganized: avionics and aeromechanics were merged into a single project, to improve both projects' integrative response. Kobi Reiner was appointed chief systems engineer of the extended project, and, effectively, of the entire program.

In this new position, did the weight of his management work increase at the expense of his engineering practice?

Kobi Reiner: "As chief systems engineer, my job was still, first and foremost, to solve problems. What did change was how far down into the system I could descend. This was not just because of the position, but also because most of the engineering problems are found in the early stages of the project. The further along the project was, the shallower the engineering issues became. At that point, we dealt less with innovations, and more with approvals and tests. But, even as Chief Systems Engineer, the bulk of my work was still engineering related. I did, however, sense that I had much more responsibility, and of course, I met with higher ranking representatives of the client. Another part of my job was coordinating the processes of structuring the new missile's support and maintenance systems. This included writing the technical documents and establishing the training frameworks."

Problem Solving

Solving the problems that come up in the course of the work is one of the systems engineer's cardinal tasks in every stage of the project. As the development advances, changes bear heavier consequences. The most difficult problems arise during the convergence processes before testing. These processes include, among other things, the planning of the right combinations and integrations for the performance of additional tests during the convergence, intended to reduce the chances of failure during the main trial. Some areas, such as flight control, are impossible to test in the early phases and must wait until after the missile has been "launched."

Kobi Reiner describes the process of handling a problem that had come up in the later stages of the project, very close to the day of the trial: "There are times when the interim tests fail to detect a faulty system, and the problem is only discovered during the final test, right before the scheduled time of the trial. Keeping up with deadlines is very important, as is meeting performance requirements.

Often, in these situations, the problem can be located and a solution presented quickly, but in some cases, the leader and his people announce that they cannot resolve the problem in time. What then? Do we enter the process of studying the problem and fixing it, knowing that the analysis alone should last three weeks, not to mention the repair process, which might take another two months?

The answer is to find a bypass. We do not resolve the problem, but search for another solution that would circumvent it for the purpose of the trial, allowing us to meet its objectives. After the trial, the full-scale solution is applied, to be tested in the next system trial. Sometimes, in these situations, there is no choice but to sacrifice one of the trial's objectives (this is, of course, done with the buyer's approval).

This example illustrates the fact that a chief systems engineer needs to be constantly making decisions. One who is unable to make decisions cannot serve as a project's systems engineer.

We had quite a few such dilemmas in our own development program. For example, we had to converge the work of three leaders for an integrative software approval (even if each leader separately approves the software developed by him, it does not guarantee that the three will play together), and one of them turned out to be one week late. As chief systems engineer, I was faced with the following dilemma: either to delay the whole process by one week, or improvise a solution that would move the process forward and minimize any damages caused by the delay. I solved the problem by asking the leader who was off-schedule to write a program that simulated his computer, in one day. That way, we could keep going without his software, 'pretending' his computer was there."

Constraints such as these can also create management problems, such as those that stem from the project's organizational structure. For instance, the systems engineer puts pressure on the leaders to keep up with the schedule. If a delay, like the one in the aforementioned example, occurs, and is not resolved quickly and efficiently enough, a situation might form, where the other two teams are idle. The leader of one of those teams might complain to his professional superior at the engineering array, saying that he had been stressed for nothing, and had nothing to do. In more severe cases, the team that accomplished its mission and is waiting for the late team to catch up is given work on another project in the meantime. It then becomes much more difficult to bring that team back into the fold of the delayed project.

Kobi Reiner: "Even back when I was a disciplinary systems engineer, one of my main goals was to get to the end of a project with minimal stomach-aches on the developers' side. It is important to hear what is on their minds; it gives them a good feeling. I never gave the developers instructions without going through the leader, but I approached them to hear the goings-on."

The Completion of the Project and Further Insights

Kobi Reiner had left his position as chief systems engineer, before the development of the missile was completed, and was appointed another high-ranking position at Rafael: chief systems engineer of the engineering array. Next are some of his other insights on systems engineering.

On Being the Chief Systems Engineer of a Project, and Being the Chief Systems Engineer of the Entire Engineering Array

In Rafael, chief systems engineers are assigned not only to projects, but to the engineering array as a whole as well. The company's engineering array includes specialized professional departments: an electronics department, a software department, an electro-optics department; each with its own chief systems engineer. The chief systems engineer of the engineering array is not a chief systems engineer on the operational level. His occupation is more focused on forward thinking – on improving technologies and optimizing development processes. Project schedules tend to be extremely demanding. A project's chief systems engineer is therefore more of an operational engineering manager. The chief systems engineer of the engineering division, on the other hand, is more focused on the professional, process-related aspects.

Kobi Reiner: "I am an aeronautical engineer by trade, but my job is to be the chief systems engineer of a division that includes many engineering departments, other than aeronautics. My business card says 'Senior Systems Engineer.'"

On the Essence and Evolution of Systems Engineering

Systems engineering is connecting components together, in order to receive new functionalities. You take a body, an engine, a navigation system, a homing head – things that have no relation to one another – and you stitch them together to make them function as a missile.

Systems engineering includes management elements within it. This is why, for a systems engineer, the transition to management is easy. If a systems engineer tends to manage anyway, the transition will be natural for him. At Rafael, it is a common course of development.

There is a range of different systems engineers. Some are architecture gurus, who do not go into the integrations that bring the system to flight condition. They make only conceptual stitches, coming up with new algorithms or new technological concepts. Other systems engineers define processes and combine abilities up to the trial phases. On the project level, too, there are chief systems engineers who are nearer to management, and others who feel more at home with engineering.

Today, everyone wants to complete many projects on a short schedule and slim budget, but without sacrificing quality. The multitudes of projects, on the one hand, and the limited pool of systems engineers, on the other, create a situation where the systems engineering in a project is "thin."

One of the main problems is that not enough effort is invested into planning ahead. The heavy operational loads leave no time for people to read the materials. These are the situations where systems engineering needs to understand the benefits of advance planning, and use it to prevent failures.

On the Uniqueness of Systems Engineering in Aeronautics

Aeronautics studies are unique, because they teach many different subjects, but perceive the airborne platform as a super-system. It is, essentially, a multidisciplinary field.

There is a kind of systems engineering within aeronautics that handles constrains other engineering disciplines are unable to meet. An aircraft needs to be safe and, at the same time, able to fly – these are two opposites. Make a perfectly safe aircraft, and it will never fly. Make an aircraft that flies, but is not safe enough; and you might not land. For this reason, the aeronautics field has long since handled systems engineering problems, that other engineering disciplines have taken long years to resolve. Aeronautics experts have these solution methods built into their thought processes, while other disciplines are still struggling with ingraining them into their methodologies.

Nevertheless, in the software field, the dominant approach is the exact opposite. Software specialists are normally not interested in the ultimate goal. They have always said: "Define the algorithm for me. I don't care whether it makes the wing move or not." They see everything through the software; physical issues do not concern them.

Some algorithm experts are systems engineers, and some are not (because work on the architecture starts even as early as at this point). A flight control specialist writes the algorithm, but he should also care which hardware it will run on. When he writes the algorithm, he should perceive the required performance and account for the environment, in which the algorithm will be realized.

Many systems engineers come from the control field. This is also due to the nature of the discipline. In aeronautics, control is more systemic than in other fields. Integration is, by definition, what it does; whether you call it that or not, this is systems engineering.

3.1.5 "COMPLEX MEGA-SYSTEMS THAT CANNOT BE SUPERVISED"

An Interview with Hillary Sillitto

One of the main forces that drove the development of systems engineering was the development of technological abilities that, with the help of software, allowed for the creation of more and more complex technological systems. This created a need for a technological professional responsible for integrating the subsystems that form the complex system.

Technological systems are an integral part of the modern world. They provide a variety of services and integrate themselves into larger systems, some of which include non-technological components as well. Thus, super-systems are created, which cannot be controlled because they are intertwined with human systems.

The development and consequences of this phenomenon, and the place of systems engineering within it, were the subject of our conversation with a systems engineer, who is also a senior executive in a multinational technological company.

Personal Background

After graduating university with a degree in physics in the 1970s, Hillary Sillitto was hired to work for Ferranti in Edinburgh, a company that developed systems for the aviation industry. He had decided to study physics because he thought that field would provide him with the best foundation for his future career. According to him, physics lays down the foundation for seeing the commonalities between different kinds of systems and different types of problems.

His first position at the company, where he continued to work for 18 years, was as an optics engineer in a laser systems group.

Hillary Sillitto: "At the time, lasers were a relatively new technology. I joined the company about 15 years after these systems had first been implemented by the Americans in Vietnam. The IRF ordered laser systems from the company. A new department was founded, and young engineers, I among them, were recruited into it."

How does a physicist find himself among engineers?

Hillary Sillitto: "Most engineering disciplines rely on physics. At its core, optical engineering, lasers included, is physics. An optical engineer must understand the physics of optical systems. We had engineers from different disciplines at the department. Electronics engineers and mechanical engineers handled the structures of detectors and receivers, while physicists worked as optical and laser engineers."

He adds: "Not all physicists are theoretical. There is a spectrum ranging from the theoretical to the applied. The gap between applied physics and engineering is small, so, in effect, we were optical engineers or laser engineers. For instance, my job was to design the optical parts of the system, like the telescope that had to produce the structure of the beam."

While he continued on his career path at the company, Hillary Sillitto also studied for a master's degree in applied optics. He worked on a number of projects in the area, as an optical engineer on some, and was responsible for a wider scope of systems on others. Even as he performed these tasks, he continued being an optical engineer, but, in hindsight, these tasks possessed clear systems engineering characteristics. Gradually, he became a systems engineer in effect, a definition he embraces to this day. As he sees it, this way of development is a product of his natural tendency to provide system wide solutions to the needs of different types of clients.

At the time, which was during the 1980s, Ferranty saw systems engineering as a collection of work methods, rather than a profession or a job title on its own. It took years of working for the company before the word "systems" found its way into Sillitto's job description, when he was appointed Chief System Concept Manager.

In 1993, Hillary Sillitto changed jobs and began working for Thales Electronics – UK, an international company specializing in the aviation defense and security field, and employing roughly 70,000 employees worldwide (with its headquarters located in France).

Sillitto: "Recruited mostly due to my knowledge of systems, I joined the electronics division. At the time, the company wanted to provide a proper response to the market changes. Up until then, it had been a supplier of optical systems. However, there was a lot of competition in that area, so the management wanted to rise to a higher place in the supply chain, by being able to provide our clients with more complex systems (systems that include several integrated subsystems). These systems had higher added value. Aircraft had become more and more complex, packed with more electronics. We saw that clients were looking for ways to integrate different products, and the company was preparing to respond that need."

After 5 years of working for Thales, his job description was formally changed to systems engineer. He was appointed chief systems engineer and became the manager of a group of multidisciplinary experts that provided support for a number of the company's projects. At the same time, he was also part of a team that developed systems engineering methodologies for the company.

Around the time of the interview, in the year 2013, Sillitto was promoted to Thales Fellow (a senior position in the professional hierarchy) – a sort of organizational consultant who travels around the world and helps the managers of different important company projects to adopt more systematic ways of thinking. Up until that time, he had been the head of systems engineering for Thales UK.

Hillary Sillitto: "In that position, I saw myself as a systems engineer in a management position. Alongside my managerial work, I also performed technical surveys of projects, so I was responsible for tools and processes as well. I tried to allocate as much time as I could to the projects, to professional work, to thinking about critical problems and ways of solving them. It was a combination of management and engineering. Like sometimes telling people not to run too fast until they fully understood the issues they're dealing with."

The Increasing Complexity and Open-ness of Systems

Hillary Sillitto points out a phenomenon that has a significant effect on how today's works is run. It is a broad view that goes beyond the technological world, although technology greatly impacts the occurrence of the phenomenon itself. It is multi-systemic, the ever-growing, ever-accelerating emergence of interdependent super-systems.

In this chapter, we will present the main principles of this phenomenon, as perceived by Hillary Sillitto and based on his experience.

Hillary Sillitto: "Most problems that are found in complex systems are caused by people, rather than technical failures. We understand technology far better than we understand the way people behave. We understand the engineering aspects, but not the socio-technical aspects, because at no point in history have we continually adopted so many new technologies so quickly." He agrees with the statement that systems are becoming more and more complex, among other reasons, simply because they can be developed (not because they are needed, but that is a discussion on a different topic, which our book does not address – the authors).

Hillary Sillitto: "We connect systems together, creating mega-systems, because they allow us to do things we could not do otherwise, to solve problems that cannot be solved otherwise. They allow us to do things better, or to develop new business models and create new opportunities, like the internet has."

There is a "super problem" that stems from the formation of such enormous systems: the large number of risks these systems entail. When designing complex systems, the thinkers and planners see the opportunities and chances, but are not always able to assess the risks and try to minimize them early, at the planning and design phase.

Hillary Sillitto demonstrates: "One of the main reasons for the financial crisis of 2008 was the ability to transfer funds across the world very quickly, or, one of the biggest problems of the internet is the terrorists' ability to use it to do harm. So, the more complex systems become, the more the connections between them multiply, the higher the chances that something will go wrong, be it on purpose or due to plain stupidity. Thus, the importance of the need to balance opportunities and risks cannot be stressed enough."

This development creates new areas of knowledge like "system science," which seek to understand the behavior of complex systems. In the past, such systems operated independently, but they grew larger and larger, and connections began to form between them. The interdependency is increasing and can no longer be fully controlled.

Hillary Sillitto: "Complex systems cannot be controlled, but only influenced (on this subject, see interview with Olivier De Weck). For instance, a connection is formed between two supersystems, where there had never been a connection before: energy and transportation. The connection between the scope of the transportation and the energy consumption affects world climate. Our increased consumption of the world's resources creates the need to seriously examine alternative ways to improve our quality of life. For example, mobile phones have had a positive effect on the quality of life in Africa. They also eliminate the need for frequent traveling. The same benefits can be obtained using communication systems. Or, take the question of water: desalination requires a lot of energy, so we must find alternative ways of doing it."

Nevertheless, how do we eliminate, or at least minimize, the risks entailed in the creation of such complex systems?

Hillary Sillitto: "In some cases, it is necessary to determine (even on legislative and regulatory levels) that a system does not exist independently, but has to contribute and integrate itself into the larger system above it. Awareness of the price of a system's existence needs to be raised, not only in financial terms, but in terms of natural resources, personal safety and more.

This means we can no longer remain satisfied with just technological personnel – engineers, mathematicians and physicists – for the planning and design of super-systems. We probably need to include other professionals, like psychologists, for instance, to help bridge the socio-technical gap."

On Systems Engineering

The aforementioned suggests that as we rise through the hierarchy of systems, at a certain point, systems engineers are no longer enough to lead the inter-systemic integration. Of course, we have yet to reach this point, as today, the field is headed by the approach that a systems engineer is, first and foremost, a technological expert, usually himself an engineer, with "systemic skills."

Hillary Sillitto: "the implementation of systems thinking goes beyond systems engineering. Systems engineering focuses on the engineering aspects, while systems thinking also touches the management and social aspects that lead the system towards success."

He presents a model that includes three levels of systems engineering:

The first level: "closed systems" – the system has clear boundaries and its people do not care for what happens outside of it;

The second level: "open systems in an open environment" – an open system, connected to its environment, which can be regarded as infinite in extent;

The third level: "open systems in a closed planet": super systems that include other systems, which, at their widest expression, include the entire world: and the environment these systems exist in is not infinite but has a hard boundary, the boundary of our planet. (For example, fishermen used to treat the sea as an infinite source of fish and an infinite sink for waste. Now they have to recognize that fish stocks are limited and pollution accumulates in the seas).

As aforesaid, the apparent question is: how relevant is systems engineering, when it comes to the third level?

In this context, Hillary Sillitto agrees that the development of complex mega-systems makes systems engineering more relevant than ever, because it includes the fundamental tools that allow us to deal with such immense complexity.

Hillary Sillitto presents another model that distinguishes between two different classifications of systems engineering. This is a model by software systems engineering expert, Prof. Dave Stupples of UCL:

The first level: Systems engineering within a discipline, like optics or software.

The second level: Interdisciplinary systems engineering working across a combination of disciplines, like mechanics, electronics, or software.

The third level: Socio-technical systems engineering – the combination of technology and the human and societal factor.

The first two levels are clearly in the systems engineer's playground. But who should be responsible for the ever growing interface between technology and society?

Hillary Sillitto admits that he does not know. He agrees with the possibility that this increasing need may bring about a new profession: "maybe there will be people who will offer a different approach and start examining pragmatic aspects. Maybe they will be called 'systemists.' This level is taking systems engineering principles beyond engineering."

He distinguishes between process-oriented systems engineers and systems engineers with systemic thinking, which transcends process-oriented thinking, and involves understanding the "whole system" and how it interacts with its environment to create the desired emergent properties.

Most engineers say: 'tell me what the problem is and I will find the solution for you'; and when they find the solution, they are satisfied. They don't care whether they've solved the right problem. This is why we need people who will define the relevance of problems. People who will examine whether the systems are still relevant as the needs of the clients change. Systems engineers must be that kind of people, but there is no uniform pattern to systems engineering.

In some companies, many people who are not systems engineers contribute to the systems work on a project. On the other hand, there are people whose job title is 'systems engineer,' but who do not think systemically at all.

We asked Hillary Sillitto, how, in his opinion, processes and systemic thinking can be combined.

He says that the major factors of the product's purpose and life cycle must be identified and focused on. For this purpose, it is important to understand the design review sequence. Understanding the purpose and life cycle of a product allows us to define what is important in each phase, which decisions to make and in what order.

At the same time, he says: "The right performance indices must be determined. It is important to decide what really makes the difference between success and failure, and knowing how to organize different types of activities in order to support them. In addition, it is important to know how to combine the knowledge of which processes to initiate and the response to the question of why we need to initiate those processes. We need to understand the details but to avoid falling into their traps (on this subject, see interview with Kobi Reiner)."

On systems engineering and intercultural differences:

Intercultural differences, whether they belong to different peoples or organizations, impact people's behavior patterns, including those of systems engineers. Hillary Sillitto finds significant differences in the implementation patterns of systems engineering. He demonstrates, using the early influences that facilitated the creation of the differences between the United Kingdom and the United States as an example: "It is customary to assume that systems engineers are technology oriented, but it is different from one country to another. In the UK, there is more awareness of the soft parts of a system, of the fact that a system is a combination of people, processes, technology and information. The US, on the other hand, is more technologically oriented and pays less attention to the human factor.

One of the reasons for this is the background for the field's development. System engineering in the UK received a push in the 1930s, due to the need for developing complex defense systems, including the radar, to defend the UK during World War II. The US, experienced a similar 'push' a decade later from the Manhattan Project (The United States' project for developing nuclear weapon during World War II – the authors). The Manhattan project was mostly technological, while the air defense battle management system that defended the British civilians, which had laid down the foundations for systems-oriented operational analysis in the UK, was a complex socio-technical system, designed to get the right information to the human decision makers who were the heart of the system."

He gives another example of the differences between countries, this time from a different angle, namely, culturally biased system considerations: "Some countries, like France, the UK, US, Israel, purchase military systems to really make use of them, to provide a real military capability, while other countries mainly want the industrial advantages they can obtain through the purchase. Meaning, the chances of them actually using an aircraft for military purposes are small. There is a much greater chance that they are interested in its technology to enhance their industrial base.

Intercultural differences are expressed, among other places, in the defense industries, the cradle of systems engineering, where the main clients' (mostly governments) general approach has changed.

In the UK, up until 20 years ago, the main risk in ordering a project lay on the government (for further discussion of the switch from 'cost plus' to 'fixed price,' see interview with Ovadia Hararri). During the 90s, the country performed an integration of defense organizations and its relationship with its suppliers underwent a significant reform. The government began transferring the responsibility to the prime contractors, who had to take some of the risks upon themselves."

This interculturality meant that in order to handle the new rules of the game, the suppliers had to formulate new work patterns. They had to deepen the collaborations between them (and bridge inter-cultural gaps, not only between countries, but between business organizations – the authors), because many complex systems involve a number of subcontractors, and if one of them fails to fulfill his responsibilities, the others (and, of course, the client) might be negatively affected as well.

A solution to this problem requires systemic thinking – but it is a business, rather than a technological issue.

According to Hillary Sillitto, a senior executive in a military technology company, such cases require the following: "The companies involved in the project need to formalize an agreement that divides both the risk share and gain share between them. Such a situation motivates the companies to work together to achieve their common goal, so that if anyone has a problem, everyone has to work as a team to solve it and achieve the goal in the most efficient way possible."

He raises another problem that stems from intercultural differences, an inside problem, typical of a multinational company: "Here in Thales, we must figure out how to bridge the cultural gaps between the different countries where we operate. In many cases, the integrator in each country offers a solution for his clients, composed of systems manufactured by us in different countries." (For further details on this subject, see interview with Gilli Fortuna).

3.2

DEVELOPMENTS IN INDUSTRY AND COMMERCE AND IN COMPLEX CIVILIAN SYSTEMS

3.2.1 "THE ABILITY TO IDENTIFY BOTTLENECKS AND ELIMINATE THEM"

An interview with Dr. Gilead (Gillie) Fortuna

If we try to define systems engineering in one, cohesive sentence, it would be the ability to apply systems thinking and multidisciplinary perspective to the management of technological systems. This suggests that systems thinking is an important element of systems engineering and one of a systems engineer's main skills.

We interviewed a high-ranking manager in Israeli industry, who had begun his career as a chemical engineer and later advanced to management positions in several different industries. This pivotal trait is the focus of this chapter; a trait important not only to systems engineers, but also to anyone interested in management.

This chapter comprised of two parts. Part one focuses on Gilead Fortuna's time as researcher and senior manager in a defense company; the defense industry being one of the first to develop systems engineering and one known for good systems thinking. Part two tells of his time in the chemicals, pharmaceuticals, and dietary supplements industries, where systems engineering is given less weight.

Managing and Engineering Complex Technological Systems, First Edition.
Avigdor Zonnenshain and Shuki Stauber.

Part one: The Experience in the Defense Industry

Gillie Fortuna studied chemical engineering in the 60s and began working for Rafael as a student in the IDF (Israeli Defense Forces) academic reserve program. As a chemical engineer, he developed propellants – the substances that generate the thrust within the missile engine system, causing it to accelerate. Being naturally curious and an inherent systems thinker, he began venturing outside the fields of chemistry and chemical engineering very early in his professional career. He did not have to go far, for Rafael's focus is not on chemistry. It did not take long for him to begin looking into missile engines – the systems where the propellants he had a part in developing were being put to use. He soon moved on to developing and managing all the systems of the missile, as well as other, innovative systems.

But why should a chemical engineer delve into the workings of a missile engine? Why did the missile developers not hand him a specification of requirements for the chemical composition of the propellants, so he could simply develop them and supply them to the developers?

Gillie Fortuna: "Nobody hands you a ready-made specification. You create it together, piece by piece. When the missile developers construct a system, they have to constantly make trade-off; we made those tradeoffs together. The mission profile requires the engine and thrusters to suit the mission parameters, some of those parameters being performance, mechanical stresses, and reliability. For example, if the engine allows the missile to maneuver close to the target, the warhead can be small, and if the engine is weaker, a larger warhead is needed.

I quickly discovered that what I was really interested in was communication with the missile developers. I did not necessarily focus on the materials (by the way, understanding the composition of the materials that serve the missile is, for all intents and purposes, also an expression of systems thinking), but on having to meet all the mission requirements."

Gillie Fortuna's testimony implies that even at that early stage in his professional progression, he had essentially stepped into the role of a systems engineer. He himself does not see it that way, though. At the beginning of the interview, he claimed to have never been a systems engineer. It is true that he was never formally defined as a systems engineer, but in practice, he behaved as one, at least some of the time. It was not long before he was promoted to management positions in Rafael. His last position there, before he left the company in the early 90s, had been VP of Marketing and Business Development. By the end of the interview, he became convinced that he had indeed been a systems engineer, mostly in his position as the head of Rafael's aeromechanics division and VP of Operations; a position where, among other things, he was placed in charge of the follow-up and coordination of most of Rafael's major projects, including such areas as resource allocation and working with the buyers.

Supportive Organizational Culture

According to Gillie Fortuna, formal, academic studies provide an infrastructure of knowledge, which is important in and of itself; but the real professionalism comes from gaining experience: "I stopped being just a chemical engineer after my first year of work, I became a technologist fairly quickly. I took part in the design of the missile engine, which meant that I had to understand mechanics and computing. I was interested in how others worked, and not confined to my specific area. Gaining experience is what made me into an aeronautics expert. In the long run, it does not matter what you studied at university anymore – what matters is what you did. Rafael's organizational culture was conducive to such phenomena, and so they occurred fairly quickly."

He explains: "Rafael placed a lot of weight on people's abilities and interests. If people were willing and able, they were allowed to move forward. Some professionals want to evolve in their own professional area, while others want to look ahead (on this subject, see also the chapter on Indigo). My nature does not allow me to narrow my point of view and not see things from a broader perspective. Rafael encouraged systemic perspectives even back in the 60s. Such organizational culture has a major effect on the development of this ability in people."

Gillie Fortuna suggests that the development of his systems thinking, one of his fundamental character traits, was enhanced by an organizational culture that encouraged the use of systemic approaches.

It should be noted that he does not claim that people who wish to continue specializing exclusively in their own professional area necessarily lack systemic perspective; only that their systemic perspective finds its expression within the framework of their chosen occupation. He gives an example about a rocket propulsion expert (by the name of M.), to illustrate this point: "M. chose to remain a propulsion expert, and yet he speaks 'systemic.' He holds a systemic dialogue with those around him: he is a consultant who knows how to speak the language of the people he consults, and is highly motivated to deepen his expertise in propulsion and contribute to complex systems by using his ability to provide professional solutions that rely on his in-depth knowledge of his field. What drove me was the desire to make a difference, and constantly do entirely new things. To an extent, it was the desire to constantly shift my point of view and see different sides of the system, while improving my ability to affect it as a whole. Fortunately, I also possessed the management skills I needed to accomplish this."

The Distinction Between Systems Thinking and Systems Management – The Question of Dependency

Gillie Fortuna distinguishes between systems thinking and systems management: "Systems thinking has been with me all my life, from a very young age. Systemic management, however, is a different matter: I only reached it in 1981, when I headed the aeromechanics division, which employed approximately 1,500 people. When we were given the task of developing Israel's first satellite, I was put in charge of an entire, complex system for the first time. It is true that in my previous position, as head of department, I also managed a system, but not one so complex and innovative.

Naturally, I appointed an excellent project manager, who later became an excellent expert in space related fields, and remained one for many years."

He explains: "The subsystems within a system are interdependent. A subsystem in a product, such as a missile, is fairly well-defined, and the interdependency between its components is under your control. As head of department, almost 90% of all resources were in my hands. The system was, therefore, not a complex one."

If follows that in a very large, complex system, the dependency on external factors is far greater. Consequently, the manager's level of control is lessened, and he has to face more constraints, being part of a larger system.

Gillie Fortuna presents two examples of the systemic dependency issue, from two different periods of his time in Rafael:

The first example refers to his time as head of research department, when he handled the field of rocket materials (in the 70s): "We developed a system of deception rockets for the Israeli Navy. The navy had successfully tested the system, and was satisfied with the results. The development of the system and combat doctrine ended in 1971. At the time, Rafael had been only a developer, not a producer; we outsourced the production to Soltam. Two years later, the 1973 Arab–Israeli War broke out, the deception rocket system proved its effectiveness, and the navy decided it needed more such systems. They approached Soltam, and discovered that they had many unusable systems lying around in storage. It turned out Rafael had set a very tight tolerance range (the maximum permitted deviation from the desired dimensions – the authors). Soltam was having trouble meeting the high standard, and failed to keep up the scheduled production pace. The company never bothered to ask Rafael to increase the tolerances. The disqualified systems just lay about in their warehouses, waiting for solutions.

We were called in to address the problem, and because, in times of war, no one has a mind to 'play around with tolerances,' within days, we had turned hundreds of systems usable and made a significant contribution to wartime production.

The systemic lesson here is that the developer cannot provide efficient production specifications, if it has never encountered production processes. Therefore, bringing production and development closer together provides vital advantages.

This story also helped lead Rafael to decide to start producing its own products. It changed the perception that the distinction between organizations in the defense industry should not be between developers and producers. The new approach stated that the differences between organizations are in the types of systems they specialize in, regardless of whether they develop or produce them, or both."

The second example is taken from his time as the head of the aeromechanics division (during the 80s): "As part of the development of a powerful laser, able to hit flying targets, Rafael was tasked with the development of the laser beam, while another industry was put in charge of developing the beam's control array. The control system was very important because for the beam to be effective, it had to be focused on the target long enough to generate the heat that would damage it.

Rafael successfully produced the beam, but the project was discontinued by us, because the other industry was unable to make the control system reliable enough.

This is an example of a system, where you do your part well, but your systemic perspective leads you to decide to invest no more of your time in the project, because it cannot be completed. In retrospect, it is clear that the project had been ahead of its time."

Identifying bottlenecks:

According to Gillie Fortuna, one of the most important skills for those who see the world systemically is the ability to identify and address bottlenecks: "It is an art, the importance of which cannot be overstated, because one cannot keep investing in system upgrades indefinitely – it is too expensive. So, instead of upgrading the system, you upgrade the bottlenecks. The ability to identify them and find ways to open them up is 'the name of the game' in today's global business competition."

Part two: The Experience in the Pharmaceuticals and Chemicals Industries

After leaving Rafael, Gillie Fortuna filled senior management positions in two companies in the chemical industry – his original field of specialization. From 1995 to 2001, he served as the CEO of TAMI, the R&D institute of a large group of companies. Later, from 2001 to 2009, he served as VP of Business Development and Operations in Teva.

He relates a systemic problem he had come across in TAMI: "TAMI developed some very special things, but very few of them were actually implemented in the companies of the group. This was because the development engineers in TAMI were not well-liked by the other companies, who perceived them as pretentious. TAMI's personnel, on their part, disdained the executives of companies who were unwisely refusing to put the wonderful products they had developed to use. Moreover, the transition from development to production was too vague. Yet another problem was that TAMI was managed by engineers and chemists who focused on the success of their development projects – their measurement of success was in their laboratories, or at most, in the semi-industrial facility, and the client company's ability to implement the development was not accounted for in their judgment of their own achievements.

When I had arrived at the company, we changed that attitude. Thus, for instance, when we offered the CEO of Rotem Deshanim (one of the companies in the group) to use a certain development of ours, he said he had a different need and asked that we propose a solution that would allow him to turn a contaminated acid they had produced large amounts of into a food quality acid, regardless of the fact that we had already developed a process for producing food quality acid directly from raw phosphate rocks. To resolve this, Rotem and TAMI engineers sat down and developed a solution together, collaborating from the very beginning of the research and development process. Changing the measurement of success from the success of a development project to its successful implementation by the client company was, to me, an application of systemic perspective."

Systems engineering in the chemicals industry:

Many of the systems engineers we have encountered in this book were integrated into development projects. TAMI, however, had no systems engineers, despite being a research institute. Gillie Fortuna explains: "Systems engineering must not have been needed. Systems engineering is important in a system with numerous components that require tradeoffs to be made. In the chemicals industry, most of the systemic view stems from the need for optimization between several products, some of which are beneficial, while others are attached as part of the process. There are not as many alternatives as there are in aeronautical systems. It is possible to examine alternatives, considering the purity of the material and the cost of the product, but it does not compare to the complexity and high level of the alternative examination process required to launch complex airborne systems into the air. In the chemicals industry, the final test is the application of the development to competitive, economic production, which integrates innovation into the chemistry, the process, the product, and the way the product fits in with the needs of the clients. In the end, this, too, necessitates a systemic view of all the development and economic production capabilities, but at a different and at a lower level of complexity."

Systems engineering in the pharmaceuticals industry.

Gillie Fortuna: "In the field of pharmaceuticals, the development of the active ingredient that facilitates the correct treatment of the disorder is in itself a highly complex process, because it requires the developers to understand the workings of the human body, identify the cause of the disorder, and then derive the treatment options that use the active chemical ingredient. Once the concept's medical potential and feasibility are proven, the system that delivers the drug into the body needs to be developed as well: the active ingredient needs to get to the right place, at the optimal dosage. The next step is guaranteeing minimum side effects and risks to other systems in the patient's body.

I had the unique honor of taking part in a beautiful joining of Rafael's systems and the world of pharmaceuticals, thanks to a systems engineer at Rafael. In 1991, as Chairman of The Civilian Business Development Committee, which we had founded, I approved the first 100 thousand dollar investment in the attempt to realize his idea and develop feasibility for a guided pill, used for diagnosing disorders in the small intestine, a project which later became a highly successful commercial system, (a company by the name of Given Imaging). In this industry, the development period from the concept stage to its full realization is even longer than that of airborne systems, and some of the processes along the way necessitate the application of a systemic approach, in order to reach successful business realization."

Systems engineering in the dietary supplements industry.

Gillie Fortuna: "After I had left Teva, I worked part time as Senior VP in two dietary supplement companies (Gadot Biochemical Industries and Frutarom), and today I am a board member in a very smart dietary supplement company (Enzymotec) that has also expanded into the pharmaceuticals market. In these industries, tradeoffs are constantly made between the effects of the supplement's ingredients, the studies that try to understand the contribution of the active ingredients to the consumers' health, and the attempts to hold clinical trials at various levels of complexity. In actuality, it seems that the more importance is given to the required approval and the potential contribution, the more complex the process of proving it becomes, though it never reaches the complexity of pharmaceutical development."

The Question of Dependency and Risk Management

The previously discussed question of dependency, in the context of controlling resources in complex projects, rises to a new strategic level when brought up in the context of Teva Pharmaceutical Industries – a global company that operates in different countries, both as producer and as marketer of its own products.

Gillie Fortuna: "In Teva, the operations in each country are treated as a separate business unit. Financial and specialization related considerations have led Teva's main management to the conclusion that it is best to concentrate the global production of each product in one place. This means that, for instance, the US marketer has to sell a lot of products produced outside the US, by another business unit. This structure creates high levels of interdependency. The US marketer needs to supply products at very tight schedules and at short notices, but he depends on the goods being shipped from Israel, Hungary or India. This means that the managers of Teva make commitments without having complete control over whether or not they can be fulfilled (unlike the division head at Rafael, who commits, knowing that he has complete control of the production resources).

Complex systems, like that of Teva, are characterized by a lot of co-dependency between subsystems. Factors beyond the control of the sellers, for instance, depend on priority orders, when production resources are limited. Specifically, the representation in China, under my leadership, sold a drug to the Chinese, local market, but the factory that produced in Hungary prioritized the US market, because it was four times as profitable, rendering us unable to realize the potential we had created by marketing our product in China.

The ability to manage client commitments without controlling the resources entails a lot of risk management that only a fine-tuned cooperation between all the involved factors can achieve."

He gives another example from his time in Enzymotec that demonstrates the part risk management plays in a company's systemic perspective: "Enzymotec's business strategy shows advanced systems thinking. It focuses on lipids and develops three families of products: mother's milk, special dietary supplements and medical food. Consequently, difficulties in one market are not enough to cause significant damage to the company. It is rare to find a company that develops three lines of products, all based on the same technology – that is smart risk management."

Further Insights on Systems Engineering

On Strategic Planning

- Other chapters in this book have already addressed the planning component of systems engineering and systems thinking. In the case of Gillie Fortuna, it is a central ingredient in his approach to management: "Wherever I went, I used strategic planning, which really is just an organized thought process. It is defining the strengths and weaknesses, threats and opportunities, the vision – where you want to go – choosing paths of action and managing the risks they entail. Strategic planning is a lot like systems engineering, you use it to see the whole of the system."

On Project Managers as Systems Engineers

- "A project manager in an organization like Rafael has to have an engineering background. A project lead who is not an administrator hires one to help him manage the budget, but the systemic decisions are ones he must understand all the way through. He has to be a technological expert; even heading a division requires in-depth technological skills. The project manager is therefore the true chief systems engineer. The person defined as the project's chief systems engineer is a kind of deputy of the project manager, who also dedicates some of his time to administrative work."
- Gillie Fortuna further adds: "In my experience, I agree with the (authors') statement that, effectively, there are two chief systems engineers in a project, and one of them, as the deputy, dedicates part of his time to administration."

On the Duality of the Systemic Perspective

- A good systems engineer has to see both ways. He must understand the subsystems and their limitations; this is a critical component of systems engineering and of a systemic view. This is why it is not enough to see things from the top down, you have to be able to see bottom up as well. As the CEO of US company, Dow Chemical once told me: "A manager must maneuver between the need to focus and make operative decisions on the single system level (*zoom in*), and the ability to adopt a broad perspective of the consequences of the realization of the vision, and the path that leads to it (*zoom out*)."

3.2.2 "WELL-ORGANIZED WORK IS ALWAYS NEEDED; THE PROBLEM IS PEOPLE DON'T ALWAYS WANT TO MAKE THE EFFORT"

An interview with Boaz Dovrin

It is often said that in its early days, systems engineering evolved at an accelerated pace in the defense, aviation, and space industries. In these large and complex systems, the need to think systemically and facilitate an educated integration of technological systems was especially urgent; more so, because in addition to the technological factors, the systems had to take into account economic considerations and the human needs of large teams. In time, the recognition of the necessity of systems engineering found its way into other technologically oriented industries as well.

This chapter describes the professional development of a senior systems engineer, who had started out in the defense industry, and then transitioned into an industry, where systems engineering was less advanced – the medical equipment industry. It also addresses Boaz Dovrin's views and standpoints on systems engineering as a whole and about the systems engineer profession in particular.

Professional Development as a Systems Engineer

Defense, Military, and Civilian Background. After graduating from his studies of aeronautics at the Technion, in 1989, as part of the IDF's academic reserve program, Boaz Dovrin was enlisted into the IDF and served in the Intelligence Corps' SUAV unit. As a technical officer, he began to act as a systems engineer, unawares. Only later, in 1996, when he was discharged and joined Elbit, did he knowingly become a systems engineer, for this time, the term was in his job title.

Boaz Dovrin: "A SUAV is a system that includes not only the hull of the craft, the systems installed on it and their replacement parts; but also the training of its operators and the management of the committee of inquiry that investigates the reason a SUAV crashed. I was never trained as a systems engineer. Some systems engineers are born this way, systemic by nature. I am one such person. I look at things from more than just one angle. Still, training and experience have helped me hone my skills."

Elbit wanted to recruit a systems engineer to work in the communications field. The wanted ad listed aeronautical engineering training and a military background among the desired skills. Dovrin applied for the job and got it. From the outset, in his very first position, he had to face issues of a markedly systemic nature that concerned the human–technology interface: "A system was developed for transmitting telegrams onto airplanes. This was a collaborative project between three different companies and the Israeli Air Force. The system was not working as it should, and all three companies – Elbit, Elta and Tadiran – were casting the blame at one another. A few months after I had joined the development teams, the system began to work.

This had a number of causes, the first of which was the work method. It is important to work methodically, and not lock on to the easiest solution right away. If a problem is searched for in an orderly fashion, chances are it will be found. Even if something does not work out, it is important to keep going and check everything all the way through. Never compromise and say 'oh well, these are the circumstances ... '

The second cause, I think, is the fact that I had arrived from the outside, and had not been involved in the development up to that point. When people are in a predicament, they become irrational, their emotions run high, and they tend to blame those around them for past failures. Joint meetings gathered to deal with the problem can easily blow up, and unexpectedly stray from the technical level. Another helpful fact was my balanced, impartial perspective. I did not immediately assume Elbit was in the right. Rather, I arrived with the desire to solve the problem."

During his 11 years in Elbit, Dovrin had evolved into a senior systems engineer. In 2007, he left the company and joined the Israeli subsidiary of 'Philips Medical Systems' – a company engaged in the field of medical equipment. Philips' Israeli branch develops and produces CT scanners.[1]

[1]The CT scanner is an advanced imaging device, where the X-ray tube and detectors are spun around, allowing the body of the subject to be scanned from all sides, generating a 360° image.

Systems Engineering at Philips Medical Systems

Systems Engineering and Organizational Culture. Based in Cleveland, United States, Philips Medical Systems employs approximately 500 people, over 100 of which are stationed at the Haifa development center. Even as he interviewed for the position at Philips, the major differences between the industry he had left and the industry he was about to join, when concerned with systems engineering, soon became painfully obvious: "I understood from the questions they had asked me (in the job interview) that they did not know what systems engineering was. They were basic questions, completely out of place for someone who had arrived from Elbit or Rafael. Questions like 'what's a design review?' or 'where does one submit requirements documents, and what is to be done with them?'

Up until 15 years ago, some of the world's industries were not even aware systems engineering existed. The head of the electronics division at Philips at the time was himself a systems engineer who also arrived from Elbit, and he understood that Philips needed systems engineering. Nevertheless, it took him several years to convince everyone that the discipline was necessary for the company. There was no systems engineering in the organization, and it did not miss it. For instance, the need for systems engineers never came up in the executive training courses.

The gaps between Phillips and Elbit were so large that I could not understand how their projects worked; how they were able to manage multiple projects without synchronizing their resources. I told my boss in Cleveland that I thought we should work differently. We should prepare an organized work scheme that included all the activities we needed to perform, and the length of time each one would require. I offered to create a plan in an Excel spreadsheet, with data, like the number of people needed, the estimated time it would take to perform each task and so on. He was excited and said 'Do this for all our people in Haifa.'

It was difficult, like tilting at windmills, because most people said 'nobody is forcing you to act this way, so why invest all this time and effort? They argued that in many cases, the investment was wasted, 'because after months of work, my project is suddenly changed,' or 'the contents of the project will change ten times and there is no point in bothering with Excel tables,' or 'I'm about to move to another position, and when I do – all the preliminary work I have done will become irrelevant.'

I disagreed. To me, it would always be relevant: 'it will serve the one who replaces you. It may be difficult now, when you do it for the first time, but later on, it will make things easier for you.' In the end, everyone needs order; the problem is that nobody wants to make the effort to achieve it."

But, if this seemed like the right way to the management, why were the employees not simply, instructed to adopt "systemic" work patterns? "It was not ingrained in the company's organizational culture, and so it was not enough for a certain manager to believe in systems engineering. It does not work like that. There was an engineering manager in the company, who saw systems engineering as 'his baby.' Every conversation with him began with how important systems engineering was. However, not everybody agreed with him.

We invested a lot of energy, as systems engineers and as managers, into selling the idea that systems engineering was needed. And still, there were those, who had been with the company for 30 years, and they told us: 'when I am going to plan my next DMS (the detector array of the CT scanner, see below – the authors), I won't need a systems engineer, because I have an engineer who has been working with me for the past 20 years, and he will write the requirements. I don't need some outsider, who came here only 4 years ago. What does he know?' Changing an organization takes time. It doesn't happen in a day.

Another difficulty in convincing people to accept the change stems from the fact that, perhaps, at the level of the organizational unit, the use of systems engineering really is not always feasible. But there is no doubt it benefits the organization as a whole, because, when you work in chaos, you never know what your people are doing. This is especially true for an organization spread across several places around the world. Sometimes, some of those places are seemingly working on the same things, but effectively, each one is developing something else entirely, without you knowing about it – several teams work on the same task and there is no synchronization whatsoever between them.

Philips used a system, where if a major problem came up, they would recruit a large team to resolve it. But this method created other problems that often went unnoticed at the time, because the consequences of bringing all those people together to solve the current problem were not thought through. Nobody could say what the meaning of such a move would be.

They lacked the maturity that comes with decades of understanding; the ability to realize that systems engineering is the way to go. Even if you try to instill systemic work methods in the organization, every problem, even a minor one, can be disruptive. For instance, you bring in a systems engineer who is not good enough, and when people see his mediocre performance, they come to you and say 'there you go – systems engineering is not worth much'."

Nevertheless, Philips Medical Systems is a successful company, and its product is among the leaders of the global market. So, perhaps it is possible to succeed without adopting systems engineering work patterns.

Boaz Dovrin: "Philips Haifa (originally Elscint) was the company that originally developed the Slice CT Scanner (an imaging technology based on scanning 'slices' of tissue in the subject's body – the authors). Today, it is ranked fourth, and might soon descend to the fifth place. It still employs very talented people, and even if you were to tie up each one's arm and foot, they would still be talented and do great things.

That might be part of the problem. Extraordinarily talented people often resist methodologies. They feel hindered by them; they need to be creative, and systems engineering is, by nature, methodical. So perhaps those geniuses, the physicists of Philips, should not have to work with the methodology.

Philips will continue to be successful, because it has good people, but I fear its market share will continue to decrease in size, because the lack of order creates a surplus of manpower, and that hurts the company's competitiveness."

The management of Philips recognized this risk and brought in new managers, who are now trying to dramatically change the company's conduct. Some of those planned changes are not suitable for a company like Philips and are actually hurting it. On the other hand, some changes match the methodologies that existed in Elbit and include such practices as preparing detailed Excel spreadsheets of all the resources needed for each project; or viewing the company's manpower holistically, thus allowing the managers to made educated decisions regarding employee mobilization and planning ahead; or preparing high-level design discussions and not moving on to the realization phase until the detailed design is complete. Additionally, the betterment of the systems engineering team and the recruitment of engineers with methodological experience have allowed the company to finally begin closing the gaps. Eventually, the time and effort paid off, and Philips Medical Systems began to be convinced in the necessity of systems engineering.

Boaz Dovrin elaborates: "When I had first come to the company, there were 4–5 systems engineers in Haifa, and another 3 in Cleveland. Today, the company employs about 30 systems engineers. The field is definitely growing. Changes in organizational culture are also evident. In the past, systems engineering was perceived as something bad. If someone was offered to become a systems engineer, he would recoil. Today, people, even physicists (a physicist is considered to be the most desirable development position in Philips), are lining up to become systems engineers. They are beginning to understand the importance of it; it just takes time. It isn't easy to persuade an entire organization to switch to different work methods."

A Problem and Its Solution. When Dovrin first came to Philips, he worked on a development project in the DMS array (Digital Measurement System) – the CT scanner's detector system. The system was being developed by teams in Israel and in the United States. The DMS is a complex system (it contains 365 thousand detectors) that requires a very high level of accuracy to operate. Even the smallest signal disruptions can distort the resulting image. For this reason, the system has to be virtually noise-free.

Dovrin tells us of a problem that had come up in this context and of the way it was eventually resolved: "It is very important for the detectors in the scanner to operate at a constant temperature, otherwise the readings are distorted. In the new scanner, the temperature within the DMS would fluctuate; its regulation mechanism did not work. In previous scanners, this problem had been resolved by placing a small heater inside the DMS, with fans next to it, to disperse the air. This way, a constant temperature was maintained. In the new scanner, a new solution was implemented, which, as aforesaid, did not work.

After repeated testing, we found an air leak in the ventilation duct. In the new scanner, the ventilation method had been changed. The small fans that served different parts of the system (the power supply, the X-ray tube, and the detector array) were replaced with a single, more powerful fan, which was meant to serve all the installed systems at once. This was an example of the phenomenon referred to as 'let's reinvent the wheel' – a result of wanting innovation. Add the fact that the development was worked on by teams based in different countries (Israel and the United States) with a major time difference and competition between them, and coordination problems are a reasonable outcome. The questions that need to be asked at the beginning are not always asked.

Historically, there were two companies, each developing its own CT system: one in Haifa and another in Cleveland. After the two were merged, residual friction prevented full cooperation, creating competition and more than a few disruptions in the work process. Each team was developing a different part, thinking that the parts would interface perfectly; all because there was nobody who could see the whole of the system.

The amount of air that was to reach to DMS had been defined during the system planning phase, but the actual amount did not meet that requirement. The Cleveland-based team responsible for air flow had been presented with the requirement concerning the necessary level of ventilation, and sent their acknowledgement to the team in Haifa. This, however, was where the dialogue had ended, for no one made sure the requirement was met as promised, and, as we later learned, it had, indeed, not been."

Dovrin led the project toward the resolution of the problem, using the systems engineering toolset (which was presented earlier, pertaining to Elbit Systems): good interpersonal conduct and a systematic approach to technological failures: "First and foremost, it was important that I talk to everyone. I avoided the question of who was to blame for the problem, and focused on brainstorming for a solution. This required everyone to pitch in. We went over the system methodically, step by step. For instance, we noticed that we were having difficulties examining a rapidly rotating system, so we detached it and put it on a table. We decided we would not move forward until we made it work on the table.

Once the leak and air supply problems were resolved, the technical failure was fixed, and the system began to work at a constant temperature; we even improved the temperature control algorithm, to boot."

Managing a Group of Systems Engineers. As he rose through the company hierarchy, Dovrin was appointed manager of the Haifa systems engineering group and then promoted yet again – to global manager of the company's system engineering. In this demanding position, he not only had to run the systems engineering, but also had to continue to ingrain the understanding that it was a necessary field: "Seeing as Philips had no systems engineering culture, some of its engineers knew nothing of work methods, and had to be taught how to work correctly. For example, they needed to be told how to properly write the requirements, and how to present them to the people developing the products – the software and electronics engineers. I had to review the work of those who wrote the requirements, and inspect the way the document was submitted.

Aside from that, there was no organized methodology for integration. In Elbit, the person in charge of integration was a systems engineer. Not so in Philips: integration was carried out by testing groups, usually comprised of electronics engineers and practical engineers, because they were more familiar with the system and knew how to examine the hardware during the integration process. But the electronics engineers mostly looked at how the hardware functioned. Once they heard there was to be software as well, they stepped aside. Organizational changes have since been implemented: the hardware testers' team was made part of a systems engineering department, and today, the tests are performed from a more holistic perspective. To support this process, a 'test plan' document is prepared, listing the subject of each test, the person or persons responsible for each stage of the testing, the expected results, and timetables."

The Do It Yourself (DIY) House Project

After about 4 years at Philips, Dovrin left to found his own startup company, convinced that his systems engineering skills would help him on his new venture, just as they had once helped him build his own house: "I decided to manage the construction of my own house, not just because of financial considerations, but because I found the idea interesting. This kind of curiosity is one of the defining character traits of a systems engineer. I began the project by teaching myself about the subject, and formulating a requirements document. My wife and I defined what we wanted, then sat down and reviewed the document with an architect. Next, we presented each of the tradesmen we hired with a copy of the document. Many couples who take on projects like this start with a humble home of 4 rooms, and end up with a palace of 8. They go into a store to choose ceramic tiles, and leave with thousands of shekels worth of marble. In our case, the construction was completed one month ahead of schedule, and the cost was lower than we had planned. This was a highly irregular outcome for a project of this sort."

Insights on Systems Engineering

On the Qualities of a Systems Engineer

- To be a good systems engineer, one needs to understand both the system and the engineering. Many people have a systemic perspective, but trade-offs and other actions that characterize systems engineering are not enough. The engineering part, the ability to closely examine the smallest of engineering details, even those outside one's original field of study, is just as important. If one employs systemic sensibility, but does not design a control system, because he does not know how, or because he finds it uninteresting, then that person is not a systems engineer. If an engineer wants to design a system but is unwilling, or does not know how to prepare a requirements document, then he is not a systems engineer.
- Generally speaking, systems engineering is a work method that allows the integration of different disciplines, for the purpose of producing a complex output. But the fact that somebody studied these methods does not make that person a systems engineer. One needs to have the right qualities and the basic perception, as they are more important than the knowledge. Take someone who knows nothing about the discipline, but possesses a natural curiosity and systems thinking ability; it will take time for him to learn, but when he does, he will get results.

- Suppose I interview someone for a job, who really likes working on printed circuit boards. It is the area where he gained most of his experience, and it is what he does for a living; that is fine, but he cannot be a systems engineer. If, however, he is versed in many different areas, that means he was curious enough about them to go and study them; meaning he has the fundamental traits of a systems engineer.
- As a systems engineer, I can easily enter any field. In Elbit, for example, I was once asked to clarify the company's new pension arrangement to other people. Engineers were complaining about receiving pension reports and not knowing what they meant. My boss thought I could prepare clear explanatory sheets, after I met with the pension company, and gain an in-depth understanding of the subject. So, I did. For three months, I attended regular meetings with the insurance company representatives, and together, we wrote clear explanatory sheets, which the insurance company could use to communicate not only with Elbit, but also with other companies as well.
- One of the most prominent traits of a good systems engineer is the ability to *visualize the end result from the start*; knowing where to go, and aim for that direction from the very beginning of the project. A systems engineer who works this way usually accomplishes his missions quickly and efficiently.

On Systems Engineering and Creativity

- In principle, methodical work can resolve problems, but working methodically without taking shortcuts is only possible in an ideal world. The reality of it is that we live in a world where time and money are limited resources, so one cannot try everything. Therefore, one has to be creative. If you take the road you believe in, you will get results faster. Of course, this way, you might miss out on some even better results, because you have not tried all the options – a "good enough" solution is not necessarily the best solution.

Planning and Organizational Culture

- Organizations should adopt a planning-oriented organizational culture. If a business is disorganized, it is only a matter of time before it fails. The larger the organization, the more crucial this principle; it is less important for smaller organizations.

3.2.3 "MANAGEMENT-ORIENTED SYSTEMS ENGINEERS ALSO SEE THE BUSINESS ASPECTS"

An interview with Alon Gazit, Erez Heisdorf, and Benjie Rom

Is "systems engineer" a profession or a position? In many cases, a profession accompanies a person throughout his life; it becomes an inseparable part of his identity, the very spinal column of his career. A position, on the other hand, is just one of several stages in the progression of one's career, which lasts a definite amount of time.

We have examined this question through the standpoints and experience of three managers in a high-tech company that operates in the printing industry. All three have begun their careers as engineers; served, for a time, as systems engineers; and then moved on the take up managerial positions. As managers, they are in charge of two types of systems engineers: those who see systems engineering as a profession, and those who see it as a position.

From Mechanical Engineer to Systems Engineer

Indigo Digital Press was founded in the late 70s as a startup company in the field of printing technologies. Its founders developed a unique, liquid type of ink (before then, all printing inks were powder based). The company made a strategic decision to develop digital color printers for the commercial and industrial sectors (rather than standard photocopiers). This meant developing a complex product, for a professional market with high entry barriers.

The first steps in this area were taken in the mid-80s. It began with a newly formed team of developers, one of the members of which was Alon Gazit, a mechanical engineer by education and, currently, the company's VP of R&D.

A developer who joins a startup company must adopt a systemic view. At this stage in a company's life cycle, there are relatively few employees, who must each perform a range of different tasks. So, Alon Gazit may have joined the company as a mechanical designer, but in effect, he says: "I did more than just mechanical design. My work included several areas, such as physics, electricity, chemistry and mechanics. Interfacing with all these disciplines had helped me develop a systemic understanding and systemic abilities. Only later, in retrospect, did I realize I had, in fact, been practicing systems engineering as well."

However, even looking back, he does not believe he had been an actual systems engineer, then. This perception extends to his next position in the company, as the leader of a team of engineers: "Management did not make the difference. The team dealt with aspects of mechanical engineering, and did not have an all inclusive view of all the different disciplines. When I was a team leader, I was not familiar with the complexities of other fields. It was far beyond me."

The next phase took place in the early 90s, when Alon Gazit served as a head of a department. The team he had led had taken part in the creation of a comprehensive system for a client, and only then was he acquainted with other disciplines for the first time; a situation that led him to become an actual systems engineer.

Alon Gazit: "Our product encompasses many disciplines, all of which eventually coalesce into ink dots on a sheet of paper: electro-optics, physics, chemistry, software, to name just a few. When we began the integration of the different fields, to make the machine work, I began to familiarize myself with them, to understand them, and, most importantly, to contribute to the development of the product as a whole."

A few more years had passed before Indigo first learned the term "systems engineer," in the mid-90s. At that time, the company was no longer a startup. New employees were recruited, some of them as systems engineers.

Alon Gazit was a systems engineer for a certain period of his professional life, but he no longer sees himself as one: "I am a development manager, who encounters many systemic dilemmas in his day-to-day work. But they are different from the dilemmas of a project's systems engineer. Today, I no longer deal with specifications that need to meet client requirements."

For example, one of the phenomena that embody the systemic perspective – seeing beyond the project level – is "commonality," a situation where a certain project's infrastructure can assist with the development of other products. Another example is when a certain project manager at Indigo, who worked on "product series 4," is required to also make use of products from "series 3." This meant that the systemic view included business considerations, budget constraints, and manpower needs as well. It was about looking not only beyond the system that was the project, but also beyond the system that was the series of products (each product being a project in its own right). According to Alon Gazit, this is a systemic perspective, but it is not systems engineering – it is management.

This approach suggests that in cases like his, a systems engineer is not a profession, but just one job title on the way up to a management position. However, as previously stated, this is not the only career path a systems engineer can take. Another path exists alongside it, and for those who take it, systems engineering becomes a profession.

The Two Development Paths of a Systems Engineer

Our interview with Alon Gazit, for which we were joined by two other senior project managers at Indigo, Erez Heisdorf and Benjie Rom, has helped clear up the distinction between the two types of systems engineers.

We shall set them apart by calling those who see systems engineering as a position "Management-Oriented Systems Engineers" and those who see it as a profession "Professional Systems Engineers." The three interviewees of this chapter are, naturally, of the former category. Today, as managers, they operate systems engineers of both types.

Erez Heisdorf: "Some systems engineers want to manage other people, to organize; they have the personalities of leaders. In contrast, there are those who wish to delve deeper into their fields and mature as professionals. They are not interested in managing people; they wish to focus on the technology."

Benjie Rom: "Management-oriented systems engineers have the ability to take a broad view of the situation. Compared to them, professional systems engineers see less of the edges. They certainly possess a broad perspective, as they must, but they do not see all the business angles. They focus more on the technical side, while management-oriented systems engineers are able to see the product not only through their own eyes, but also through the client's as well."

Alon Gazit: "Management-oriented systems engineers are not afraid to confront 'neighboring' factors, and this sometimes leads to conflicts. Factors like marketing, service, operations, and even matrix bodies [note: matrix bodies are the Indigo term for professional units that support the project (see, for comparison, the engineering units at Rafael, mentioned in the interview with Kobi Reiner)]. They have leadership abilities and communication skills, and they are more willing to compromise – they are potential managers. Professional systems engineers are more solid, more perfectionists. They enjoy dealing with technology, but not the small, technological details – technology in a wider sense. This is why most of them are not inventors. These are mostly found in the matrix bodies."

Alon Gazit believes that the system engineer position is a critical step in the development of anyone who wishes to eventually manage multidisciplinary systems. When asked about the abilities a management-oriented systems engineer needs, he replies with an example: "Suppose there is a multidisciplinary problem that needs solving, and it is not yet possible to even define where it originates, and, consequently, who should be handling it. This is the type of problem a good systems engineer needs to be able to solve, by combining leadership and analytical skills. In a case like this, a professional systems engineer would find the problem more difficult, because, although he has the necessary analytical skills, he will be limited by his lacking leadership ability."

A Systems Engineer at Indigo

The way systems engineering develops within an organization is influenced by the nature of the organization, its organizational culture, and the fields it is engaged in, among other things.

Having joined Indigo as a project manager, after serving as a systems engineer in Elop, an Elbit Systems company (for more on systems engineering at Elbit, see interviews with Mimi Timnat and Boaz Dovrin), Benjie Rom relates his first-hand experience of the differences between systems engineering in Indigo and systems engineering in other organizations. According to him, there is a vast difference between the companies, in terms of their systems engineers' responsibilities: "Systems engineering at Elop is a very clearly defined discipline. The main difference is that the systems engineers there are responsible for designing parts of the system, while here, at Indigo, they have no part in the design. Here, a systems engineer can share his experience with the designers, or take the group in a certain direction, because he has the necessary experience, but he does not actually design; that is done by the matrix bodies. The main reason for this is the complexity of Indigo's products, which necessitates the placement of a systems engineer in each technological group, thus reducing the need for the systems engineers to deal with the project's more technological components."

Alon Gazit provides another comparison: this time, with other technological bodies within Indigo's parent company, HP (which acquired Indigo in 2002): "In the print field, there are similarities between job titles here and at HP. But, in most cases, a systems engineer at HP has less room for growth. The reason for this is Indigo's organizational structure and technology, which force good systems engineers to interface with many more disciplines than the average systems engineer at HP."

He demonstrates: "In the late 80s, HP developed the inkjet technology, which included two main areas of development: the ink and printheads, and the printer itself. Each field was assigned to a different organization within HP, which had its own separate systems engineers. In our company, the technology that facilitates the interaction between the ink and the machine is much more complex, and forces the systems engineer to deal with a much wider range of disciplines. So, in principle, there are similarities, but the levels of complexity are different."

Erez Heisdorf expands on the subject of complexity, by demonstrating: "One of the subsystems of a printer is the feeder, which allows both sides of the sheet to be printed on, and then proceeds to stack the sheets, after the printing process is complete. The group tasked with the development of this subsystem has its own, internal, systems engineer, because the system is a combination of mechanics, electronics and software, and someone needs to manage its systemic aspects. The machine has five or six such subsystems, and they, in turn, must be synchronized with each other. We call the systems engineer in charge of this subsystem a 'functional systems engineer'."

At Indigo, the tasks a systems engineer receives match his abilities and traits. It follows that there is more than one type of systems engineer in this company.

Alon Gazit: "To me, a project's systems engineer is the highest technical authority. I can give an example of such a person. He was the right hand of the program manager. He was also in charge of the distribution of the specification's components between the different units, the distribution of the budget to the various disciplines, in addition to integration and validation [validation is a process that makes sure the system meets the specified requirements]. Alongside that, there were other systems engineers who worked underneath him, and dealt with the details.

The matrix bodies also have systems engineers. For instance, the software body has a systems engineer who manages the software field, and underneath him, there are systems engineers who have to do their best to define the software before it is written, and coordinate between the different software teams that write it. So, there indeed are different types of systems engineers."

The Growth and Development of a Systems Engineer at Indigo

During the first years of Indigo's existence, systems engineers emerged from within the ranks of its employees. Some of them, like Alon Gazit, have since become the executive spine of the R&D division. The source of this phenomenon is discussed in the beginning of this chapter: in a small organization, there is a high chance for systems engineers to handle a number of different areas and be intimately familiar with the company's range of activities. Hence, a large part of the systems engineers who evolved at Indigo belong to the "management-oriented" type."

But, as the company grew, so did the difficulty of developing new systems engineers out of existing employees, and the need arose for recruiting systems engineers who had grown and gained experience elsewhere. A considerable part of these systems engineers could be classified as "professional systems engineers."

Alon Gazit is of the opinion that a major cause of this is the company's size. Size necessitates specialization, making the employees of a company focus on narrower, more specific fields, and reducing the chance of their learning a wide range of technologies. One possibility of bringing about such development, besides recruiting systems engineers, who have already gained experience in other organizations, is sending employees to be trained outside of Indigo. Yet, these courses mostly help broaden their attendants' horizons and foster new thought patterns. Naturally, they are unable to meet a company's concrete needs. These needs can be met in on-job training, as well as internal training courses.

Other chapters in this book have already addressed the question of which base discipline systems engineers grow out of. Some have argued that the answer depends on the core disciplines the organization deals with. It appears that this is true for Indigo, where a considerable part of the systems engineers began their careers as mechanical engineers.

Alon Gazit: "Traditionally, in our company, systems engineers start out as mechanical engineers. This is a direct result of how Indigo was established. The company had set out with many mechanical engineers, because the machine has a lot of 'metal.' The integration process is also carried out by 'metal.' But systems engineering is more than just hardware; almost every field of mechanical engineering is present here, from thermodynamics to optics, to materials engineering."

The people of Indigo believe that systems engineers should, ideally, be engineers by trade, although they can also be physicists. In their experience, the engineering background of systems engineers is usually in either mechanical or electronics engineering (they perceive aeronautical engineering as a subdiscipline of mechanical engineering).

And what about software engineers? Alon Gazit: "There are systems engineers with that background, too, but they usually prefer to take up handling the systemic aspects of software."

Further Insights on Systems Engineering

On the fundamental traits of a good systems engineer

Benjie Rom:

- Curiosity and the desire to learn different fields, rather than specialize in one discipline.
- The ability to make decisions and distinguish the essential from the nonessential. When employing a broad perspective, this becomes an everyday need, because there is never much room to maneuver, as there are not only technological constraints to work within, but also a demanding schedule and limited resources.
- The ability to delve into the details and analyze. This is true for both professional and management-oriented systems engineers, with one difference: depth. Still, both must have an analytical mind.

Erez Heisdorf:

- Good systems engineers need to be able to work on two depth levels simultaneously. They must be able to go into the small details when they meet with the work teams, and be able to see things from very high up at the same time. People cannot always "live" on these two levels; it is a unique ability.
- Common sense.
- The ability to make decisions.

3.2.4 "OPTIMIZATION BY THE TOP RANKS"

An interview with Dr. Amir Ziv-Av

Systems engineering is a new discipline that still seeks to define its place and interrelations with other disciplines and fields that exist alongside it or overlap with its areas of activity. One such area is optimization, the purpose of which is to find an optimal value for functions, under given constraints.

We have stated repeatedly that systems engineering emerged as a result of the increasing complexity of technological systems; a complexity enhanced by the development of the software field and its integration in technological systems. But technology, complex though it may be, is usually only one part of larger, more intricate systems that assist modern human society in its day-to-day conduct. One of the more obvious examples of such complexity can be found in transportation systems.

This chapter deals with the combination of these two areas, namely, optimization and complex technological projects, and their affinity toward systems engineering. It revolves around an interview with Dr. Amir Ziv-Av, Chief Scientist at The Ministry of Transport, who also wrote his doctoral dissertation on systems optimization.

Personal Background

Having graduated high school with a practical engineering diploma in mechanical engineering, Amir Ziv-Av's enlistment with the IDF Ordnance Corps was only natural. After being discharged from his military service with the rank of an officer, he decided to further his study of mechanical engineering and applied to Tel Aviv University. Even as a student, he worked in his area of expertise, providing planning services to private companies and to the development units of the IDF's Combat Engineering Corps. After receiving his Bachelor's degree, and having simultaneously completed part of his Master's degree studies, he received an offer to return to his old unit. As a young Major, he headed the Department of Mechanical Development, with a team of 12 engineers under his command. About two years later (in 1979), after completing a (short) command and headquarters course, he was transferred to the headquarters of the Engineering Corps, to found the corps' own development branch. At the same time, he was also busy with projects related to the Merkava Main Battle Tank.

Amir Ziv-Av: "During that time, I began effectively practicing systems engineering, as I formulated the methodology for the analysis and development of the process of breaking through minefields. I used the Air Force's operations research branch, because The Engineering Corps still spoke the traditional language of "mines per frontline meter," while I wanted to speak in terms of optimizing the resources used by an attacking force attempting to overpower a defending force in a barricaded position. Clearly, I was leaving the boundaries of my area of specialization – mechanical engineering. My work in the development branch enhanced my systemic perspective, particularly the ability to define and analyze a system. It was my first practice of systems engineering. When you are the head of an entire corps' development branch, you do not actually do any developing, rather, you define and accompany the development process. The means at the disposal of any corps are multidisciplinary in nature."

In the early 80s, after approximately six years of standing military service, Amir Ziv-Av left the IDF and completed his Master's degree in mechanical engineering. During the 80s, he served as Head of Opto-Mechanical Development in Optrotech, a company that developed automated, optical systems for inspecting printed circuit boards. He did not arrive at the position as a systems engineer, for the term was not widely known at the time. Nonetheless, his career path was paved by "systemic" traits.

Amir Ziv-Av: "I had not come from the hi-tech industry, nor was my education of great relevance to the position. For the most part, Optrotech hired me because of my systemic perspective. Most of the people who reviewed my candidacy were physicists.

Mechanical engineering is not the ideal background for a systems engineer, and the ability to practice systemic subjects depends more on the person than on the discipline he specialized in. A systems engineer does not always have to be an engineer, and the higher you climb to look at the system from a bird's eye view (and the further away you get from technological issues), the less necessary is it for you to be an engineer.

I am also not afraid of venturing into unfamiliar areas, because when you look at everything from up high, you apply the same principles to all fields. I learned the principles of optics, asked questions and received all the help I needed in my search for optimization. The questions that arise in optics are no different than in other fields: 'which trade-off should I choose?,' 'what is the level of mechanical accuracy needed here?,' 'what extent of control or accuracy do I need?,' or 'what is the risk level posed by this new concept?'."

Systems Engineering and Optimization

Amir Ziv-Av sees a close connection between optimization and systems engineering, because one of systems engineering's main goals is optimization. He believes that one of the core characteristics of systems engineering – a holistic, all-inclusive perspective – stands at the heart of what optimization really means: "viewing the system as a whole, not as one discipline or another, but as an ensemble of economic, operational and technological components. The essence of the connection between optimization and systems engineering is the development of optimal concepts for correct integration of all the technologies at our disposal. A 'product' is an answer to a collection of differently weighed objectives, and at the heart of its development process stands the task of maximizing the target function. In the end, to win the competition over the heart of the client, one must have a relative advantage, and how does one obtain that? By doing more with fewer resources; that is the bottom line."

Which brings us to the importance of intuition: "One shouldn't force quantification on everything. Some complex situations include so many parameters, that weighing each one is impossible; other times, there is missing data. This is when intuition comes into play."

A well-devised strategy can greatly enhance the ability to make the right decisions, even under conditions of uncertainty: "The product concept is its strategy, and the product tactic is its details. If the strategy is good, mistakes can be corrected, even if the tactics are wrong. Take, for example, development. Being an iterative process (a process that repeats itself in order to examine a range of different situations. If errors are discovered, they are fixed, and the process begins again – the authors), a good development strategy keeps the developers on the right track. A bad strategy, on the other hand, is very difficult to correct, no matter how good the tactics."

One of the best examples of this, according to him, is the 2006 Lebanon War, where an incorrect strategy led to failure and to the eventual appointment of a committee of inquiry. This happened despite the impressive heroics seen in the field and the good field rank performance at the battalion and company levels (to this we, the authors, add that had the strategy been better, the soldiers and commanders in the field may have had fewer heroic acts forced upon them): "They were unsuccessful in changing the general direction. The same holds for product development. When the general direction is the right one, corrections can be made en route; but, when the project is heading the wrong way, there is nothing to be done."

After his time in Optrotech, Amir Ziv-Av transferred to Keter Plastic, where he served as Head of Development and management member. His experience there taught him, once again, of the importance of good strategy: "The CEO and owner of the company had a clear strategy in mind. He defined the direction the company was headed. At the time, the company's scope of sales had been roughly 50 million dollars a year; today, it is close to one billion. Everything he had said he was about to do, he did: he decided what to produce and what not to produce, who to compete with and who to cooperate with, where to establish local factories, who to ship to, when to launch which product and what risks to take in the process, who to copy from and when to take initiative, and at which point in the factory's growth it needed to be split in two.

The internal workings of the company were another matter. There were various conflicts and differences, such as competition between two internal elements in production. But when the strategy is so good, these little things are not enough to hinder it."

The striving toward optimization entails two main principles: robustness and simplicity.

Amir Ziv-Av: "When you set off on the road, especially if it is a long or tricky one, there is a lot of uncertainty. This is why *robustness* is such an important component in optimization – it is the ability to aim for a system that is insensitive to changes in the operating point. This is achieved with wide design margins (referring to reserves, not to marginal situations) and modularity, so that when something new comes up, it can be integrated into the system with no difficulty. This way, even if, ten years from now, the system is not in its prime, it will be in the right environment. And if there is a problem, it will either be easy to fix, or, its consequences will be possible live with."

Amir Ziv-Av then talks about the Trans-Israel Highway (also known as "Highway 6") as a good example of the robustness principle. When the highway was being designed, the traffic load and number of lanes needed were impossible to predict. So, only two lanes were paved in each direction. After the highway proved to be surprisingly popular among commuters, it was decided to add another lane. Seeing as this was taken into account during the planning stage, the addition necessitated no changes to existing infrastructure: "We did not have to touch a single interchange connecting the highway to the lateral roads, not one bridge, street light or drainage system; even the guard rails were left almost entirely untouched. It was a very robust solution."

Simplicity means action is not always necessary. Sometimes, the right thing to do is to do nothing, making inaction the optimal solution.

For example, the changes in Dizengoff Square: "Dizengoff Square was a beautiful, safe plaza with good air circulation – a pleasant urban environment on all accounts. They took a good thing, invested tens of millions of shekels in it, and turned it into a bad thing. Now, the streets are congested, and the atmosphere is unpleasant."

Amir Ziv-Av gives two examples of strategic failures in areas nearing his current occupation: the aviation and automotive industries.

On unnecessary developments in aviation, the failure of which, could have been avoided beforehand:"The Concorde airliner was a technological masterpiece and, at the same time, an economic failure. The fact that the investment would not be returned could have been predicted. It was a known fact that the price of a seat on a passenger airline was derived from the cost, and ranged around 300 thousand dollars, independent of the airline model. A seat on the Concorde cost around one million dollars. No airline ticket can justify such a price.

To return the investment entailed in the development of a passenger airliner (and we know that the investment in the Concorde was not far off that of the Jumbo jet), one needs to sell between 200 and 300 units. The Concorde sold only 15 – a massive economic failure, while the Jumbo sold over 2000 – a huge success.

The Concorde's advantage of speed was only relevant for the time spent flying across The Atlantic, seeing as the extra time spent by the passengers at the airport before they boarded the plane and after they got off it remained unchanged. The downside – three hours spent on a crowded plane – did not justify the difference in ticket prices.

A similar phenomenon occurred with the Airbus A380, a giant jetliner that required adjustments to be made in many airports to accommodate it. Why did the developers assume the airport authorities would agree to change their runways and terminals just for them? Moreover, the process of shipping the enormous parts of the body of the plane from the subcontractors to the assembly factory was extremely complicated. The fact that so far, only a few dozen places were supplied is a testimony to the project's failure. Each year, the cash-flow is delayed by many more billions of dollars – a financial disaster."

On redundant mergers and acquisitions in the automotive industry: "Some of the mergers that took place in the automotive industry were unnecessary, and ended in major losses. Mercedes, a profitable, stable, focused company, joined with disorganized and inefficient Kreisler. The endeavor ended with the two companies splitting up and suffering billions in damages in the process. Ford bought and then sold Jaguar and Volvo, losing money on both deals. GM began producing the utter nonsense called the Hammer, only to sell it later at a loss."

To the question of how, in spite of everything, these fiascos come about, Amir Ziv-Av responds with one sentence: "The megalomania of executives makes them lose focus, and focus is almost always a condition for success."

Systems Engineering, Optimization, and Transportation Systems

One of the most prominent expressions of the mutual affinity between systems and larger systems can be found in the content worlds Amir Ziv-Av inhabits today. As the Chief Scientist at the Israel Ministry of Transport, National Infrastructure and Road Safety, he is in charge of formulating the strategy of transportation systems.

For instance: when planning the layout of a road, the considerations include such parameters as road length, which needs to be minimized, so that the road takes up as little area as possible; and the radii of turns, which should be maximized to raise safety levels. The planners strive for making the road as inherently safe as it can be, to lessen the need for additional safety means, such as guard rails. Other considerations include minimizing incident sensitivity (among other reasons, to avoid a situation where an accident causes a roadblock; a major problem even for short lengths of time). All these parameters then need to be considered from a systemic perspective, so as to reach the optimal result.

The rate of mileage increase in Israel is three times the rate of increase in road area. The authorities are attempting to close this ever-increasing gap in two ways: the first, lowering mileage by transitioning to available and inviting public transit, and the second, increasing the efficiency of land infrastructures. Another helpful means would be changing behavior patterns by encouraging people to start their workdays at different hours of the day or working from home, but it lies outside the jurisdiction of the Ministry of Transport.

Infrastructure efficiency necessitates optimization, which relies upon systems engineering. This is similar to aerial transport, where the planner can, for instance, decide that in three months, on a certain day, at a certain time, a certain flight will pass over Greece. But such accurate control is impossible in land transportation. So, according to Amir Ziv-Av, the authorities settle for dealing only with the national transportation network and targeting congested areas: "We cannot manage the single car traveling from Be'er Sheva to Arad, but we can manage heavily congested regions like the Tel Aviv, Haifa, and Jerusalem Metropolitan Areas. We need to plan the system so that, for example, a train does not leave five minutes before the bus arrives at the train station. To prevent that, we need to manage traffic signal timing or interchanging lanes, when most people travel in a certain direction; all while giving priority to public transit and emergency service vehicles, as needed."

Today, traffic data is available to any driver, so they can do their own, individual optimization. The massive use of smartphone devices allows this level of data accessibility. The pedestrian or driver will simply state his destination, and the device will present him with the optimal route, suited for his individual needs. Technology will also work against such nuisances as queues forming at the entrances and exits to and from parking lots, by using automatic billing, a technology that already exists and is used today. An automatic billing system, like the one in use on Highway 6, can be installed at parking lot entrances or exits.

An example concerning the use of cameras as a means of enforcement.

Amir Ziv-av: "Transportation in Israel is considered safe by international standards. One of the main reasons for this is the substantial improvement of infrastructures (grade separation, guard rails, and roundabouts). The use of road safety cameras for speed limit, red light, and center line enforcement is another important factor. Studies have shown that placing cameras on road systems can significantly lower accident rates. In France, adding 3,600 cameras to support enforcement has lowered the traffic related death rate by 50% in ten years. An equivalent effort in Israel would require 1,500 cameras. At present, however, there are a mere few dozens. Expert opinions and relevant authorities (like the traffic police and the Road Safety Authority) believe this investment to be worthwhile, and human benefits aside, it is also feasible economically. Nevertheless, implementation has been very slow."

He gives another example, this time, of non-systems thinking: "One of the problems with implementing traffic control programs is that the technological systems are not integrated. It is a 'Tower of Babel' type of situation. The traffic light control center of the Tel Aviv municipality does not serve nearby cities, whereas the traffic itself belongs to a single, metropolitan urban entity. The decision to use a certain technology is mostly driven by the results of tenders, each of which took place at a different time, under different circumstances. On a national level, there is no optimal, all-inclusive, systemic-technological perspective. In other words, there is no systems engineering. For example, the Tel Aviv Municipal Control Center may find it easier to manage all fifteen cities of the Tel Aviv Metropolitan Area, than to wear itself out trying to interface with them, as it does today."

But Amir Ziv-Av is optimistic: "I am currently leading the ministry in this very direction – toward comprehensive systemic optimization, on a national level. All in all, people want things to change for the better, so long as the change does not come at their expense. Still, they recognize the need for integration and coordination on the road towards finding a systemic solution that overcomes constraints; for example, by establishing a metropolitan authority with representatives from each of the member municipalities."

Further Insights on Systems Engineering

On the evolution of systems engineering

– "Today, most products are interdisciplinary, and so the need for systems engineering is on the rise. 30 years ago, for instance, cars were, for the most part, mechanical products; so were jumbo jets. Today, cars are equipped with computers and communication systems, even the mechanism that opens the window has some small processor in it – everything is mixed together. The Smartphone is at the forefront of technological, software and communications knowledge. But if it can fall on the floor, survive a shock of 500 Gs and not break, it means its mechanical engineering is also cutting edge."

On the essence of systems engineering and its affinity for optimization

– "Systems engineering means zooming out. Not dealing with the molecules, the 'micro,' but with the macro. If you zoom in, you will see the molecules. A systems engineer does the opposite – he zooms out, decides on the various disciplines that, together, form the solution, and defines the interfaces between them. Systems engineering is optimization of the highest order. It places a lot of weight on operations research and entails many legal, as well as economic considerations."

On a systems engineer's professional background

– "A systems engineer doesn't necessarily have to be an engineer. The central trait of a systems engineer is a comprehensive view, an ability based on the systems engineer's personal skills, rather than his area of study. Of course, this depends on which level of the system the engineer is stationed in – the higher the level, the less engineering skill is needed. The more you zoom out, the less important an engineering background becomes."

On engineering and systems engineering studies

– "Decades ago, engineering studies were four years long. They are still four years long, today. How can this be? Knowledge has increased a thousand fold – they say it doubles every few years. The answer is that luckily, base disciplines – physics and mathematics – change at a much slower pace. These are the subjects that develop skills. Those who study them receive the tools for learning and understanding the other disciplines. Mechanics, for example, includes several fundamental physical principles, and the rest is mathematical developments. If you studied mathematics and physics, but know nothing about heat transfer; when you open a book on heat transfer, you will be able to understand it, even if you are not an energy engineer. Perhaps, in the distant past, students learned a large part of the mechanical knowledge in existence during their first degree studies; while today, they learn only a tiny bit of it. But if the engineer has the ability to learn, that does not matter."

3.3

THE INFLUENCE OF THE ACCELERATED PROGRESS IN THE COMPUTING WORLD

3.3.1 "WHEN A CRITICAL MASS OF PROCESSES AND METHODS IS FORMED, A NEW PROFESSION IS BORN"

An Interview with Henry Broodney

One of the dilemmas systems engineers have to face is the question of the reciprocal relations between the systems engineering profession and basic engineering disciplines. After all, at its core, systems engineering is a methodology that ties the different engineering fields together. In order for it to do this successfully, its importance and necessity must be recognized by the engineers in those classical fields.

Technion Prof. Aviv Rosen (see Section 3.4.4) is of the opinion that the fact that systems engineers concern themselves mostly with the links between the components of a system, rather than with its professional engineering level, might be the cause of a rift between them and the engineers in the field. It follows that if systems engineering does not make a real connection between itself and the other engineering fields, it will find it difficult to evolve further. Rosen believes engineers expect to see "the link between systems engineering and physics and mathematics." One of the main ways to achieve this objective is the development of computerized systems engineering tools. In recent years, computer companies like IBM have begun developing exactly such tools, in collaboration with systems engineers representing the various industries, who help the company by pointing out their industries' specific needs. This process

Managing and Engineering Complex Technological Systems, First Edition.
Avigdor Zonnenshain and Shuki Stauber.

is a step toward the development of software tools that would provide the industries with the solutions they require.

This chapter discusses the happenings in this field through the story of an electronics engineer who had been employed by the defense and aviation industries, then, later, became a systems engineer, and currently manages the "Systems Engineering Technologies Unit" of the IBM R&D Center in Haifa.

From Electronics Engineer to Systems Engineer

After graduating from his studies at The Technion's School of Electrical Engineering within the framework of the IDF's academic reserve program, Henry Broodney joined the Israeli Air Force's EW (Electronic Warfare) array as a project officer: "As I handled the maintenance and upgrades of EW systems, my first encounter with systems engineering happened early in my career. I had to understand what client requirements were, learn how to translate the pilots' dreams into something practical. I worked with the operating companies and with software specialists from another unit that provided us with its services. Still, at the time, (the early 2000s), I did not think of myself as a systems engineer, only as a specialist officer in a technical field."

Broodney learned the trade on the job. The only formal training he received during the four years of his military service was two courses: an EW course and a Project Officers course.

In 2003, Broodney was discharged from the IDF (after receiving his MBA degree from the Technion). His experience in the military was his ticket to the electronics field in Rafael's EW department, where he held his position for a very short time. At the time, Rafael had won the right to take on a series of projects in the fast-evolving field of weapon stations (remote weapon control performed from a command and control station). Broodney was recruited into a new project in this field as a "Unit Leader in the Electronics Department," which in Rafael-speak meant he was in charge of all the electronics within that project.

There, he had to acquire new skills, both technical and system-integration related: "In the Air Force, I never had to make a device work; I only had to verify that it did. In Rafael, I was required to make sure the devices worked, formulate specifications and standards and then compare them to the requirements."

During his three and a half years in Rafael, Henry Broodney's systemic mindset gradually matured, until he finally became a full-fledged systems engineer: "I gradually began to venture into fields other than electronics, like mechanics. For example, if a box was being built to contain the electronic components, I, as the leader of the electronics units, had to instruct the one who built it. Still, I did not see myself as a systems engineer at first. I worked with the project's systems engineer, whose place was in the project's directorate, and he was the one who coordinated the project's various technological fields.

Later, when I found myself working on other projects (Henry Broodney led the electronics units in various projects during his years in Rafael), I began to realize I was no longer occupying myself exclusively with electronics, but with the entire system. In my last project in Rafael, I was still an electronics engineer, but I was a systems engineer too."

There were other factors, besides Henry Broodney's coordination efforts with other technological disciplines, which brought him to the realization that he was acting as a systems engineer: it was a time when systems engineering was slowly affirming its presence in the awareness of the industry. Rafael, too, had done much to advance the systems engineer profession within its own walls: it held internal systems engineering courses and sent its people to take part in external seminars and professional conferences. Then, there was the gleaming aura of management.

Henry Broodney: "I found the area appealing for me. Systems engineers in Rafael sit in directorates, dynamic business units that work directly with the clients. There were quite a few engineers of my age in the company (only in the fourth decade of his life, Henry Broodney is our youngest interviewee – the authors) who aimed for that position. There was a feeling in the air that the best engineers were going to become systems engineers."

In 2006, Henry Broodney's last year in Rafael, he took part in a systems engineering course, as part of his advancement within the company. At the end of the course, however, his career progression was cut short, as he decided to leave Rafael and, together with a partner, founded a start-up company in the software field. In this new initiative, he found himself with a range of new responsibilities: he managed various aspects of the emerging company's business, including fundraising, finances, recruitment and human resource management, product management, and more. The start-up proved to be a promising venture, and employed seven people at its peak, but the global economic crisis of 2008 struck a fatal blow to the dream of Henry Broodney and his partner. Financing sources ran dry, and the two were forced to end the venture. Henry Broodney, who by then had accumulated some valuable experience in both management and technology, felt his way back into the corporate world and managed to get a job offer from Soltam, a company that specialized in metalwork and ammunition. This offer was the final seal of approval for his status as a systems engineer – the head of the company's Department of Artillery offered him the position of a systems engineer in the company administration.

Henry Broodney: "Soltam was undergoing major changes at the time, as part of its recovery from a recent crisis. The position I was entering was a new one. The Head of the Administration, who had also transferred from Rafael several months earlier, offered me the job. We tried to instill systems engineering work patterns in Soltam. I also learned a new field along the way: heavy mechanical production."

During the period of his employment by Soltam, the ties between Soltam and Elbit were becoming increasingly close. The two companies collaborated on several projects, and Henry Broodney found himself working intensively with Elbit personnel. During the second year of his employment by Soltam (2010), the company was acquired by Elbit, and its people took over all the high ranking positions. Finding no career prospects in the newly reconfigured company, Henry Broodney chose to leave.

Not long after, he received a surprising job offer from the IBM R&D Lab in Haifa, to work, once again, as a systems engineer. Why did the computing giant's research laboratory need a systems engineer, whose background was in electronics and whose experience was in defense systems?

Indeed, ordinarily, IBM hardly ever needed people of Henry Broodney's professional background. However, at the time, a special opportunity arose at the company, and Henry Broodney fit into it like a glove.

Henry Broodney: "The mandate of IBM's development units is to develop innovative technologies for use in IBM's products. Around that time, IBM's subsidiary, Rational, whose product was software development tools, acquired a company called Telelogic, which specialized, among other things, in developing (software) modeling tools for use in software engineering. The Haifa Research Lab saw an opportunity, and the executive who recruited me began to form a team to take up this field. He made contact with the US Defense Advanced Research Projects Agency, the organization in charge of developing the American defense system, and they started working on a certain project. At this point, he needed a content expert who was familiar with the inner workings of the defense industry and knew how to use computerized modeling to work on a project."

Soon, other software development projects were referred to the Haifa Lab, and roughly one year later, IBM's management formally approved the establishment of a group that would develop systems engineering technologies. At the head of the group, which currently numbers 12 members, is systems engineer Henry Broodney.

The group began by developing software tools for systems engineers in the aviation industry. Later, demands began to arrive from other areas as well, such as the automotive and transportation systems industries.

Systems Engineering at IBM

IBM employs thousands of people in "systems engineer" positions worldwide. The company makes efforts to advance the systems engineering profession and organize a body of knowledge of the discipline and holds professional gatherings and seminars. Yet, Henry Broodney, who had spent much of his career in the defense industry – the cradle of systems engineering, holds that most, if not all of them, are not really systems engineers, but IT engineers who use systems engineering methodologies in their work. In other words, they are systems engineers, whose work is confined to the IT field. Additionally, most of IBM's systems engineers are not placed in the company's R&D units, but in its sales, marketing, and service units. The purpose of this strategy is to allow these units to speak the same language as the systems engineers of IBM's potential clients, those who will eventually be offered to buy IBM's computerized systems, which will help them in their work. Henry Broodney says: "Perhaps, with clients like Lockheed Martin, one does need a full-time systems engineer."

It follows that IBM will not need many systems engineers of Henry Broodney's type in the foreseeable future. It might need one systems engineer for each new field its development center decides to develop products for, because "IBM is not going to develop products, but product development tools."

The research activity of IBM's R&D labs is not pure, empirical research, but rather, applied research – the kind that aims toward the development of products, for which the industry will present a demand. However, IBM does not sit idly and wait for clients to place orders for the development of products that suit their needs (although sometimes, as foresaid, development takes place in collaboration with the client). Rather, it tries to understand where the industry is headed beforehand and develop software products to suit its future needs. This is the mission of the unit under Henry Broodney's management, centered on the specific case of developing tools to help systems engineers in various industries. This is also why he is perfect for the job. As a systems engineer who evolved in industries that are considered to be the spearhead of the systems engineer profession, Henry Broodney is an external content expert, able to help the IT giant become familiar with the needs of its potential clients in the systems engineering world.

Henry Broodney: "We mingle with the various industries, hear their problems, and then identify the ones IBM is able to offer solutions for. The next step is to find a partner, inside or outside IBM, and to start developing the solution.

Physically, the systems are similar; the difference lies in their domains. Some systems are mechanical intensive, while others are hydraulic intensive, like the equipment of an oil company. For the research teams developing the systems engineering software, the specific engineering discipline is of little importance. We do not need to know the engineering minutiae; we only need the top level. Then, we use the services of experts who usually come from our research partners, like Boeing or Daimler AG. IBM recruited me as a systems engineer who knows which products are relevant to the industry he came from."

Henry Broodney explains further: "There are two types of systems engineering, the first of which began to emerge as far back as the 60s and 70s. It includes processes and work methods for the management of projects that include three key elements: scope, schedule, and money. The second type of systems engineering deals with engineering planning and the process of designing the system itself. The first type focuses on system management, meaning what the project manager does, whereas the second type focuses on the engineering, meaning what the project's chief systems engineer does. This is the focus of my unit – technical, not management-oriented systems engineering.

The planning stage is crucial: 70 per cent of the product's costs, reliability, and performance are determined during the project's preliminary stages, when preliminary design is done right. This perception is slowly seeping into the awareness of the industry. There is an increasing demand for (computerized) tools that help design good system architecture; tools that help the developers understand the interaction between the software and the physical components it controls."

Insights on Systems Engineering

On the evolution of systems engineering

- In Roman times, it was enough for one smart man to understand how things worked. Systems were simple, back then. One talented man could see the whole picture, weigh the right considerations, and take many different aspects into account. Things worked more or less the same way until World War II, when big names like Wernher von Braun and Willy Messerschmitt were still prominent, and one man's talent could give birth to major innovations.

Starting from the mid-twentieth century, the leaps and bounds of technology have created a situation where being smart is not enough, and in order to innovate, one has to structure the effort. One also needs processes to help him think, because systems have become far too complicated for the capabilities of any one person. Teamwork has become a necessity. All of today's greatest inventions are the results of group efforts. Hence, we need processes that support group work. So it happens that, when the processes and methods reach a critical mass, a new profession is born, and, like anything new, it needs a name, a buzz word to help position it.

On the identity of systems engineering

- Systems engineering has evolved differently in each organization and has been called by many different names. Moreover, in each organization, systems engineers perceive their position differently (if that organization even has systems engineers). In Elbit, for example, there are no systems engineers; instead, there are "technical managers," a name that offers a rather accurate description of the job. There is an administration called the Systems Engineering Administration, but the people who work there are called "technical managers." In Rafael, on the other hand, there are systems engineers who are essentially technical managers.
- Most systems engineers in Rafael have backgrounds in either mechanics or electronics. If a project is not assigned a systems engineer, the electronics engineer automatically received the title, being likely perceived as the one most suitable for the job and best able to see the whole picture. In my estimate, the reason for this is the fact that systems have begun to be increasingly software-heavy, and software engineers have a good grasp of several areas, because all electronics engineering curriculums include software studies. These two factors created a situation where electronics engineers are assigned the positions of project systems engineers.
- In the automotive industry, there is no "systems engineer" position. There are, however, "architects," who essentially fill the same role and will, in time, become INCOSE members. Currently, most INCOSE members are employed in the defense and aviation industries, but more and more new members are flocking to it from other industries, as they come to realize that they are all members of the same trade. The industries may work in different ways, but the body of knowledge is the same: there is a wheel that turns and a software controller that oversees its movement.

The qualities of a good systems engineer

- Today, one person can no longer manage a large project. For example, one cannot demand that a project's chief systems engineer understand the workings of the project's financial management. He needs to recognize that there are financial considerations, or know how the design is reflected in the cost, but the costs themselves must be planned out by someone else.

– A systems engineer needs to know "enough," not "more." He needs to be able to handle many things at once, to multitask. I find it difficult to focus on something narrow and systematic. I am more holistic than analytical by nature; I am also highly intuitive. But that did not make me a systems engineer; the chain of events that introduced me to systems did. To this day, I miss delving into the inner workings of a technical system.

3.3.2 "LOOKING AT A PROBLEM FROM DIFFERENT ANGLES"

An Interview with Mimi Timnat

Systems engineering has evolved differently in every organization, in accordance with its unique needs and organizational culture. In a considerable number of cases, a system engineer's job title is not even necessarily "system engineer." The job descriptions tend to vary as well – the range of tasks a system engineer is given can change greatly from one organization to the next.

In this chapter, we will expand on the use of systems engineering in Elbit Systems, through an interview with Mimi Timnat, a high-ranking systems engineer who has filled a wide range of positions throughout her career, from her beginning as a software engineer to her current position. Among other subjects, our conversation also revolved around the evolution of software engineers into systems engineers.

From Software Engineer to Systems Engineer

Much like many other systems engineers, Mimi Timnat, currently the Head of System Engineering and Technical Management Process Improvement at Elbit Systems, had set foot on the path toward systems engineering unintentionally, as part of her natural development, guided by her personality and interests.

The transition to systems engineering drove Mimi Timnat to discover fields of engineering that have nothing to do with software, and lead the development of multidisciplinary systems, in which software plays only a minor role.

Having completed her studies in the department of Computer Science at the Technion (as part of the IDF's Academic Reserve program), she joined Elbit Systems as a software engineer and was tasked with software development. Two years later, she was promoted to Software Project Manager. In this position, she started to become acquainted with systemic issues, in addition to software-related ones.

Mimi Timnat: "Even as a Software Project Manager, I had discovered, as had many other software managers, that my duties often required me to go beyond mere software implemented solutions. We often discovered, in the midst of the software development process, that some things were difficult to implement in accordance with the predefined requirements or that the definitions themselves were not clear enough. I usually understood what was needed. On some occasions, I would not wait for the systems engineer to become available and provide me with a definition; rather, I would suggest solutions myself and then coordinate them with the systems engineer.

Back then (in the 80s), Elbit Systems had already had systems engineers who were referred to as such, but their job descriptions were rather amorphous. At the time, the practice of systems engineering was almost an art form: there were people who knew what to do, but no methodological foundation and no organized training (the field of software had very similar beginnings, by the way)."

Mimi Timnat's transition to systems engineering took place sometime later, in the 90s, and was completely unplanned and unintentional.

Mimi Timnat: "My transition to systems engineering began spontaneously, following my participation in a design review that discussed the solution to a complex problem. My part in the review was supposed to be insignificant – I was only to test whether the part that was meant to be implemented using software could indeed be accomplished. The solution presented in the review was based on the unification of partial solutions, which, together, formed a solution for the larger problem. When I saw the solution, it seemed to me that in certain situations, the all-inclusive solution would prove impossible. It was as if they required one car to be in two places at the same time. I asked whether I had understood the solution of the suspect situations correctly, and was met with silence. As it turned out, the solution really was faulty."

Shortly after, she was offered to temporarily assist Systems Engineering with resolving the problem and then return to Software. She brought up the proposal with her superior.

Mimi Timnat: "I approached my superior and told him about the offer. He said: 'If you move to Systems Engineering, I do not believe you will want to come back to Software. If you want to go back, you can. But, I think you will find the field attractive, and you will choose to stay.' It later turned out he was right."

Mimi Timnat's transition to systems engineering was not limited to solving the problem that initiated it; it opened up whole new worlds for her.

Mimi Timnat: "I joined meetings with clients, and was required to understand their expectations and look for solutions. I was exposed to engineering fields I had never dealt with, and terms I had never learned. At first, it was strange and unclear, but along the way, I asked questions, learned and began to understand many engineering subjects – far beyond mere software. I was naturally attracted to the need to understand the whole picture, and project-oriented systems engineering has given me that option."

How does a software engineer, whose knowledge base and experience are rooted in the software field, acquire the extra knowledge a systems engineer needs?

Mimi Timnat: "In any engineering discipline, the knowledge acquired in studying does not last very long. Therefore, one must always remain up to date. One of the things that can be said to the Technion's credit is that, among other things, they developed their students' independent thought and learning abilities. During the course of my work, I taught myself much about previously unfamiliar areas and used the help of others as well. In my estimation, people who find knowledge and understanding important find ways to acquire the knowledge they are missing. Even when I go to the doctor, I ask questions and try to gain a better understanding of things."

Today's sophisticated systems combine a variety of engineering areas and specializations. It is difficult to expect one person to be an expert in everything. One of the challenges of developing a complex system is handling the interactions between all the different disciplines and their specialists. Thus, Mimi Timnat holds that one of the systems engineer's most important tasks is to understand engineers from various disciplines and coordinate between them: "My growth as a systems engineer was an unplanned process, where one thing led to another. I clearly remember a discussion with many participants from various disciplines, about ways of implementing a solution for one of the issues in the project. At the beginning of the meeting, I raised several questions on subjects I was not clear on and wrote down the agreements we had reached. Later, other engineers raised other issues, seemingly irrelevant to me. There were many professional terms I did not know, I could barely understand what was being said, and I found it hard to keep summarizing the discussion. After the meeting was over, I asked one of the other engineers for help clarifying the unfamiliar subjects and checking whether I had summed them up correctly.

That meeting was just the beginning. As time passed, the subject was raised again and developed further. More details were added and improvements to the agreements we had arrived at in the preliminary discussion were suggested.

At the second discussion on the same subject, I was asked to update my earlier summary. The discussions were not easy. People from different engineering disciplines were having trouble understanding the problems and solutions of other groups. Sometimes, people used the same terms to refer to different things. Without planning to, I had become a 'translator' for the different groups (note: a similar problem is described in the interview with John Thomas).

During the course of the project, that same "protocol" had evolved into a specification, dozens of pages long, wherein were listed the details of a solution that involved several engineering disciplines. And so, to my surprise, as I was updating the "protocol," using my ability to understand the different groups and coordinate between them, I had become a knowledgeable leader on the subject, within that project."

Software Engineers as Systems Engineers

Previous chapters in this book have already addressed the subject of software engineering and its reciprocal relationship with systems engineering. Particularly, they raise the argument that software specialists tend to concern themselves with software and are less willing to deal with the other engineering disciplines. On her part, Mimi Timnat claims that such generalizations cannot be made and that these attitudes depend on the people: "Today, there are many disciplines and subdisciplines, even within the umbrella of software engineering. For instance, the expertise required for designing video games is not the same as the one needed for developing organizational information systems or communication applications. Some prefer to specialize in software, others, who have a tendency toward working with systems, enjoy working in large projects, in systems of systems, and being able to understand the whole picture. These people choose to evolve into systems engineers, because it allows them to grow laterally. This way, they familiarize themselves with many different fields, like control, communications, mechanics, optics and more – all as the project requires. Everyone finds what is right for him.

I believe it is important for a systems engineer to have a good grasp of the dominant area in the project he is developing. In software-heavy systems, software engineers have a natural advantage, but there are systems where the dominant area is not software. An example of that is robotics. In these systems, mechanical engineers and electronics engineers have an advantage, as systems engineers who also studied software, but not as thoroughly (today, this is true for all engineering disciplines). In computer science on the other hand, there is no requirement to study other disciplines, which is why, in some projects, software engineers find leading the engineering of the system difficult."

Another phenomenon presented herein, in this context, is the increasing number of software engineers found among systems engineers. Mimi Timnat agrees with the existence of this trend and explains it by saying that software is slowly taking up a central position in complex systems: "In the past, hardware components had much more dominance in systems than software components, and so a considerable part of the systems engineers rose from disciplines of relatively high technological complexity, such as electronics. Today, software takes up much more weight, causing more systems engineers to evolve from that area. The career change from software to systems engineering entails a shift of focus. The broad perspective becomes more important; the focus shifts to understanding the comprehensive solution that pertains to all disciplines, and how the system is compatible with the user; and away from the details of software implementations."

This is the place to note that Technion Prof. Aviv Rosen argues that the case of Mimi Timnat, who has achieved high-ranking positions in systems engineering, is a rarity. He stands among those who hold to the previously presented opinion, according to which the lion's share of software engineers still prefer to focus on software, and only a minority are willing to tackle other fields. He believes the reason for the growing number of systems engineers who rise from among the ranks of this group of engineers is that software engineers have greatly increased in numbers in recent years (in many projects, they constitute the largest group of engineers). He claims that the relative share of systems engineers who started out as software engineers is still fairly small.

On the Identities of the Systems Engineer and Chief Systems Engineer

Systems engineers are usually engineers by trade, but Mimi Timnat believes it is possible for systems engineers to be trained in other areas as well. "It depends on who the developers are. If most of them are engineers, and they expect a systems engineer to be well versed in the project's dominant discipline, then, naturally, he has to be an engineer. But it is possible for a developer of, for instance, a biomedical system, to be a doctor with a broad perspective and some comprehension of engineering, even if he is not an engineer per se.

Sometimes, an engineer can come from the operational side. Then, he would have an excellent grasp of client needs. I know some excellent systems engineers at Elbit Systems who rose from operational disciplines and have no engineering degree. For example, it is said that some of IAI's best systems engineers are former IAF pilots. In any event, it is important for a systems engineer to have several years of hands-on experience working on development projects."

The subject of professional background relates to the question of a chief systems engineer's suitability to the project he is meant to lead. Mimi Timnat believes the job of a project's chief systems engineer is influenced by the association between the project's technological core and the systems engineer's original discipline: "Having an understanding of a range of different fields is a fundamental condition for a systems engineer. He must be able to recognize what he knows, what he does not know, and when to consult other experts. But a chief systems engineer needs to have a very good grasp of the project's lead discipline. For example, I, as a systems engineer who started out in the software field, would find it difficult to be the chief software engineer in a project laden with mechanics or robotics. Like other disciplines, systems engineering has specializations."

This leads us to the question: if two of the most important qualities a systems engineer must possess are learning ability and a broad perspective, why would a potential chief systems engineer not learn the new area and lead the project? Why should he continue to focus on his generic area of expertise?

Mimi Timnat: "It can be done, but learning takes time. Projects usually have very tight schedules, so prior knowledge is preferable. If a chief systems engineer has gained relevant experience working on a similar project, it is best to use it. Engineers are indeed quick learners, but if there is the possibility of finding someone with relevant experience, we try to do so."

Systems Engineering at Elbit Systems

The evolution of systems engineering varies from one organization to the next. It depends on each company's needs and organizational culture. At Elbit Systems, a company that develops and produces advanced electronic and electro-optic defense systems, systems engineers focus mostly on the technical areas, but need to account for a variety of considerations, including budget, scheduling, risks, and more. Elbit's organizational culture makes all engineers consider these constraints, but the core of their work is, as aforesaid, engineering and technology oriented.

In addition to "systems engineers," Elbit also has a position called a "technical manager." Technical managers bear a general responsibility for the execution of a project's development, which they coordinate with the project manager (who, in Elbit, is known as the "Program Manager"). The program manager is more concerned with the business and contractual aspects of the project. He outranks the technical manager, but the latter is in charge of the project's technical areas. In most cases, the technical managers' occupational history includes the position of a systems engineer at Elbit. A program manager can grow out of the company's engineering areas, but may also come from a business or operational background and with no engineering experience.

Having filled the positions of Software Engineer, Software Project Engineer, Technical Manager, and Program Manager, Mimi Timnat advanced to a new, unique position within the company – Head of Systems Engineering Methodology and Development Process Characterization. This position's uniqueness stems from the fact that only a select few of the hundreds of systems engineers employed at Elbit Systems work outside the framework of a project. They are organized in a centralized body that provides services to projects and business divisions that deal with system development.

Further Insights on Systems Engineering

On the Essence of Systems Engineering

- Systems engineering is a discipline that deals with problems and solutions that combine many different fields of engineering. It includes a variety of tasks, all of them involved with the process of developing a project. These include analyzing requirements, formalizing the concept of the solution, integrating, testing, and more. Systems engineers deal with different engineering fields at different scopes, and therefore, there are different levels to their positions. Systems engineering focuses on application, rather than research. This is why academia finds it difficult to "raise" systems engineers, unfamiliar with development projects. It is difficult to explain how to handle the issue of "understanding the needs of the client" until one meets a real client who has trouble defining his need and has to do it for him.

What Makes a Good Systems Engineer?

- (Mimi Timnat relies on the results of a study on the subject performed by Prof. Moti Frank):
  - Broad perspective – "This ability is vital, and not necessarily inborn; it can be learned. One is taught to ask questions, to examine things from different angles. One is given examples of typical questions, and considerations, useful when comparing alternate solutions to different problems. This is how this ability is acquired."
  - Good self-learning ability.
  - "The ability to distinguish between what one knows and what one does not."
  - Resourcefulness and flexibility – "A systems engineer constantly encounters unexpected problems and mismatches between different engineering disciplines. During the course of the project's development, he has to deal with the reactions of developers from different fields, such as 'what you defined cannot be done' or 'these constraints cannot be met.' It is important

to constantly check whether the main requirements are met and come up with creative solutions for any surprises encountered along the way."

- Good interpersonal skills.
- Leadership – "It is important to not only find the correct solution, but also convince the involved parties to accept it."
- The ability to "move" things.
- The ability to communicate with the client and come to understand his expectations – "This means seeing things from different perspectives, being able to use other people's terms, understand their real needs, even if the client himself is having trouble defining them."
- The ability to delve into the details, without losing the broader perspective. Because "at times, only when you get to the details of the implementation, do the problems begin to show themselves."
- Common sense.
- The ability to formulate solutions in conditions of uncertainty – "Because the systemic solution concept needs to be designed in the project's early stages, when many details are still unclear. The systems engineer needs to be able to understand what can and cannot be implemented, without knowing all the details (and in larger systems, a single engineer can never know all the details)."
- A systems engineer needs to know when he understands enough, when to ask for more details, and how to get them. For instance, when a client asks for a change to be made in mid-project: at first, the systems engineer makes a preliminary assessment of the consequences of the change, in order to see whether the change is even reasonably possible. Only if he finds the change plausible, does he consult the leaders of the various disciplines for a deeper examination of the consequences of the change and the budgeting of its implementation.

Mimi Timnat summarizes: "I find that, to a great extent, systems engineering is more than just a job. It is an approach to handling and solving problems, and not only work-related ones. It encourages one to look at a problem from different angles, to ask questions and try to gain a better understanding of the problem, before making decisions and formulating solutions."

3.3.3 "VENTURING BEYOND THE CORE-SUBJECTS TO STUDY NEW AREAS"

An Interview with Harold (Bud) Lawson

One of the major factors for the constant increase in the complexity of technological systems and integrated systems is the dramatic increase in computing and communication, which has allowed for the development of impressive technological capabilities. One of the pioneers of the global computing industry, Bud Lawson, believes that Systems' Engineering evolved as a discipline intended to assist us in dealing with such complexity.

This chapter presents Bud Lawson's developmental path; a man who initially was not trained in engineering, had developed into a leading computer expert and later on into a systems' engineer. His transformation into a systems' engineer had come about with a clear awareness that is not always characteristic of computer experts, as one cannot function as a computer specialist without understanding the overall system that a computer-based solution is supposed to serve.

It's Not the Computer, It's the Application

Bud (Harold) Lawson, a computer engineer-cum-systems engineer, who is considered one of the pioneers of the global computer industry, never studied computers in an orderly fashion, since no organized studies in the field existed, naturally, in the late 1950s, the period in which he entered into this emerging industry. Born in the United States, Lawson's formal educational foundation was in mathematics and statistics, which he studied along with economics.

In 1959, Lawson began working in the new computing industry that had intrigued and excited him. He learned from experience. He worked at Remington Rand Univac and then at IBM between 1961 and 1967. Being an expert in programming, he was involved, during his first years in the company, in a series of projects dealing with the development of computer languages and compilers. At a later stage, he decided to change track and get involved in computer design, focusing upon the field of hardware.

During those years, the industry was dominated by the big computers, first and foremost of which was the IBM System/360, whose development and penetration into global markets the Company had focused its attention upon. This approach was detrimental to Lawson as an entrepreneur and developer. In 1967, he headed a small research team that examined one of the fundamental problems of the computer world – the relationship between hardware and software. But at the same time, according to him, "IBM was not open to ideas that could prepare it for present, as well as future, operations. The Company traveled along the path of the '360.' As such, even though the research had discovered ways to improve the hardware–software interface, it did not receive the attention it deserved."

Lawson left IBM, joined the academic world, and had also become a senior consultant in "Computer Engineering" – which was practically a new area of activity at the time and which did not exist in academia.

Lawson became a professor and a senior consultant in "Computer Engineering," even though he had not explicitly been trained nor worked as an engineer, in any classical sense. According to him, "the field was called 'Computer Engineering' because it dealt with the design of hardware and software systems and with defining the relations between them." He began to study and join projects in the United States, but mostly in Europe. He relocated to Sweden first as a consultant to Datasaab and joined the Linköping University faculty, where the field of Computer Engineering was becoming highly active.

We shall present two prominent projects in which he took part during those years (the 1970s), as a senior computing consultant. The systemic perspective, which lays at the foundation of systems engineering, is prominent in both. This was so, even before he saw himself as a systems engineer. Eventually, this approach, by his own definition, turned him into "a systems engineer, specializing in computer-based systems."

His first project was related to the development of a remote monitoring system for high-voltage power transmission.

Bud Lawson: "One of my students was from Barcelona. He had worked at the Enher power company that developed a remote monitoring system (tel-control system) for power distribution. To that end, they needed to construct a computer-based system. He asked me to assist them. We discovered several interesting solutions that were successfully implemented. This is an example, not only of the work done on the computer itself, but also of its applications. For me, that marked the beginning of a shift, because as I look back and examine my work, it seems that I have always worked at systems engineering. But that conceptual shift began, for me, in the mid 1970s, when I stopped focusing upon the computers themselves and started looking deeply into applications for computers."

But were you aware then, in the mid 1970s, that you were formulating an approach that was different from that of most contemporary computer specialists?

Bud Lawson: "I felt that I was doing something different, but I did not use the word 'systems engineer'. I had worked at computer engineering previously, as well. But after I had started working on the applications, I told myself that I needed to learn much more about other areas, such as power distribution. If I wanted to design a suitable solution, then I could not focus just upon understanding the computerization part, I needed to understand the area that the computer was supposed to serve."

The second project involved the development of an automatic control system for trains, the first of its kind in the world.

Bud Lawson: "In the mid 1970s, operations, which could not have been done in the past, were made possible by the microprocessors that started to appear on the global market. These microprocessors changed the economic world of hardware. Standard Radio & Telefon contracted to Swedish Rail asked me to take the lead in a development project, the first of its kind at the time (no such system was in operation anywhere in the world) – computerized monitoring of the train network, which until then had been electro-mechanic. For instance, the system would allow us to monitor the drivers' behavior. If, for example, the train driver suffered a heart attack (and such things do occur), the monitoring system will automatically stop the train; or, you could define speed limits between the signal segments. Such a system also optimizes the engine's operational performance.

The company that was put in charge of the project was as mentioned 'Standard Radio', a local subsidiary of the ITT communication corporation. When I had first arrived on the scene, they already had an active team that had constructed a simulator that included what the system needed, to their understanding. I looked at what they built and said: 'there are going to be a lot of problems, because there's a lot of 'spaghetti code' (long and complex lines of code – The authors) that are difficult to understand'. I suggested that we go back and examine the system's requirements.

A key requirement was the ability to maintain stability every 250 milliseconds, or one quarter of a second. You can do much with a computer in that sort of timeframe. I told them: 'Let's look at this differently, as if it's a clock cycle. It might be hardware, but let's look at it as if the software' was hardware. I suggested a different approach, to look at the system as a continuous system rather than as a discrete system; since the most critical variable here is time. In this way we transformed it into and far simpler solution. We built a system that had only 4,000 instructions and only 10 KBs of code. We installed it in around 1,000 train engines and it worked very well."

(Note: Looking at the system as a continuous system prevents us from focusing upon *specific* (discrete) instances, usually rare instances, which also require a solution from the system. As an analogy, we can imagine the system as focusing upon the flow of a wide river without being required to focus upon its dozens or hundreds of tributaries. The river flow is the essential point, and we focus upon it. Obviously, such a concept reduces the system's complexity and cumbersomeness – The authors).

This approach has served Lawson in additional instances as well, while allowing his skills as a systems engineer to manifest themselves. He is one who is wise enough to use a solution that has been successful in one system to seek out solutions in a different system.

For example: He was approached by Haldex contracted to Volkswagen in the 1990s, to assist them in the development of a device that could increase the fuel efficiency and safety in their 4 × 4 vehicles.

Lawson: "They encountered problems caused by their historical mechanical view of product development. To them dealing with electronics and software was a new experience.

These were basic problems of systems engineering *such as the need and usefulness of well-defined processes* that were the result, among other things, of mechanical engineers doing all the work, even though they did not understand electronics and computers. Once Haldex realized they had a problem, they approached me as a real-time computer systems expert. I developed for them a monitoring system that was similar to the system originally developed for Swedish Rail. These days, this component is installed in most 4 × 4 vehicles in the world.

In order to address the development of this automotive component composed of electronic hardware, software, and mechanical parts, I tailored a version of the ISO/IEC 12207 (Software Life Cycle Process) standard in order to include not only software, but also the electronic and mechanical hardware. This was prior to the development of the Systems Engineering standard ISO/IEC 15288 of which I became the architect."

Insights into Systems Engineering

On the essence of systems

- A system is a collection of items that creates something that possesses new functions. As a result, we sometimes witness the emergence of new patterns of behavior.

- There are hard systems, and there are soft systems. As time goes by, even my fellow engineers understand that they need to see the human side, along with the engineering\physical side. The word engineering has been used to describe soft systems: social engineering or human factor engineering. We must be aware of these aspects when we design a system. When we design an aircraft, we need to take into consideration the person intending to operate it. Even if it's a completely automated vehicle, there is still a human component.

On the complexity of systems

- Many systems suffer from over-complication, from an overabundance of unnecessary functions. This makes monitoring and controlling them difficult. The problem is especially prevalent in computer-based systems, because programs are virtual thing and can be developed endlessly. In the past, computer memory was relatively small and as such made the development complex and complicated programs impossible.
- A systems engineer is supposed to prevent, as much as possible, unnecessary complexity in the system and to deal with the problems caused by complex systems as well. On the one hand, a relatively simple system needs to be created, one that can be worked with. On the other hand, there are systemic problems that require modification and the development of features that assist in dealing with their complexity (meaning that we need to add elements to the complex systems that will help in dealing with their complexity issues – The authors).

On the essence of systems engineering

- One of the problems in systems engineering is that their design is based upon engineering systems, and not upon the broad perspective within the context of a system, which includes the people operating and using it. Systemic thinking is an area that has yet to be well defined, despite the fact that we use such systemic thinking, in practice, in our daily activities. The Academic world is currently researching the science of systems, systemic thinking, and their connection to systems engineering. But the purpose of science is to create understanding, not presenting solutions.
- Systems engineering began with people focusing upon narrow aspects of the system*s*. When we develop systems with different aspects such as mechanics, electronics, computers, and so on, it involves multiple disciplines. Achieving results – the creation of products and services – requires the creation of cooperation between them. Therein lays the crux of the problem – someone needs to bridge between them.
- In Sweden, systems engineering is not studied under this name – they use other names such as "industrial management." I myself teach in educational programs, which fit in better with programs that include systemic embedded infrastructure. These programs are not called "systems engineering" due to the division between those areas that involve engineering and those areas that do

not really involve engineering. So, even though we recognize the importance of this issue and might even pay it lip service, integration between disciplines is not so simple in reality.

On the characteristics of the job

- There are those that will say that systems engineering is both a profession and the discipline. For me, it has become a profession. I am a systems engineer that specializes in computer-based systems.
- A systems engineer does not have to be an engineer. I am not an engineer by education, but I am an engineer in practice. It is not a profession but rather a way of thinking. We are constantly finding solutions to problems.
- A systems engineer needs to be open to the study of new topics and not focus solely upon his area of expertise. Thus, I have learned much about engineering in practice.

Authors' Insights:

On the complexity of systems

- Complex systems are developed, not because the clients require all of their functions, but rather because the ability to develop them exists – Lawson concurs.

On the essence of systems engineering

- Systems engineering is a developing area that has yet to define itself – Lawson concurs.
- The professional community accepts Lawson as a systems engineer, even though he lacks an education as an engineer, because his professional background is computer sciences, which is seen, in many cases, as belong to the world of engineering.
- Despite the fact that in a variety of fields, there is a tendency to adopt the thought patterns of systems engineering, there are forces, which are very strong, which cling to the traditional engineering professions, which do not take this area of expertise for granted. To them, systems engineering looks like a discipline that goes beyond its true area of expertise.

3.3.4 "THE ABSTRACT LEVEL OF DISCUSSION IS OF GREAT VALUE"

An Interview with Sharon Shoshany Tavory

There is a unique, reciprocal relationship between the two worlds of software and systems engineering. Many claim that this relationship emerged from traditional, physical engineering. The need to plan and integrate complex, technological systems has raised the need for a design and integration toolset that has since become the infrastructure of systems engineering. Software, however, is not always perceived as part of the engineering world, mostly because of its virtual nature. Technion Prof. Aviv Rosen tells us of a conversation he had had with Amos Chorev, former president of the Technion, who had wondered: "Where is the engineering here?" In another chapter in this book, Kobi Reiner says that as a systems engineer, he had, on several occasions, encountered problems when working with software specialists, because they had been unwilling to see the limitations of the system as a whole, persistently focusing almost exclusively on their software-related tasks.

Nonetheless, software is often the factor that facilitates system integration. This means that the ability to design software bears a lot of weight in the development of systems engineering. Today, the discipline's two overarching missions, namely, preliminary system design, followed by integration of the system components, entail massive use of software-based tools. Moreover, some would claim that software engineering resembles systems engineering more than any other engineering discipline, because they both create virtual engineering environments, unlike traditional engineering disciplines, which, as we said, are anchored to the physical world.

This chapter will present the career progression of a chief systems engineer who started out as an electronics engineer, transitioned to software engineering early on, and then moved on to evolve into a chief systems engineer.

Her views and the progression of her career tell the story of the evolution of systems engineering in recent years, with an emphasis on its unique reciprocity with the world of software.

From Electronics and Software Engineer to Systems Engineer

In 1986, two years after she had begun working for Rafael, electronics engineer Sharon Shoshany Tavory was recruited to work on a large project. She was tasked with developing real-time software (a program that had to meet requirements for performance under certain, predetermined time-related constraints) for certain parts of the project.

What would an electronics engineer have to do with systems engineering?

Sharon Shoshany Tavory: "At the Department of Digital Systems, where I worked, the prevalent perception was that real-time software development should be done by electronics engineers, because software people could not see the whole picture. They only saw their own 'bits'."

In those days, Sharon Shoshany Tavory had already begun practicing systems engineering, but she only came to realize it about a decade later, when she discovered systems engineering as an emerging methodology: "At those days, the term 'systems engineering' was already in existence, but it did not bear the meaning it does today. It was usually used to refer to someone who got promoted to Assistant Project Manager. Back then, I had begun doing systems engineering work. For example, I managed interfaces and analyzed processes, but I was not called a 'systems engineer'."

Sharon Shoshany Tavory's entrance into the world of systems engineering was also influenced by her personality, as well as by her work as a software developer:

"I was not introduced to systems engineering during my academic studies at the Technion. At the time, it was not considered to be an academic discipline. I acquired my expertise in the field on the job, learning from experience. Every engineer starts his professional career as a 'screw'. He can choose to remain a screw and focus on the field of engineering he specialized in, or he can choose to look around him. I chose to look around me.

At that time, software engineering had only just begun creating tools for systemic work and systems thinking. Because software is abstract by nature, it had an especially great need for tools such as analysis of requirements or systemic abstract modeling. One can safely say that systemic models were born in the world of software."

In 1999, Sharon Shoshany Tavory, then head of Rafael's Department of Digital Systems, decided to go on sabbatical in Australia. It was there that she encountered the terminology of systems engineering in full force: "This was a much 'hyped up' subject at the time, and people began seriously considering formalizing and institutionalizing systems engineering as a discipline. One of the consequences of this was the change in the name of the institute, where I was a guest, that year. When I had arrived, it was called the 'Test and Evaluation Center,' but during my stay, it was renamed 'Systems Engineering and Evaluation Center'."

In the year 2000, Sharon Shoshany Tavory returned to Israel and was appointed Project Manager. In this position, she was faced with the varied challenges of project management for the first time, including such responsibilities as working with outside clients, like the Israeli Air Force and US aircraft manufacturer Lockheed Martin. In her previously held positions, she had served clients from within the organization, including project managers, who received professional services from her and the department she led.

Was this not too sharp a transition? Would it not have been better to undergo a gradual training course that included, for instance, an interim job as Deputy Project Manager and/or a project's Chief Systems Engineer?

Sharon Shoshany Tavory: "It is true that I had never attempted actual project management before then, but I had gained experience in similar frameworks. First, as a department manager, I, of course, had management responsibilities: I devised work schemes, negotiated with clients within Rafael, and managed employees. Second, I was no stranger to contact with external clients, either. As a department head, I met with clients who wanted to gain a deeper understanding of the specific field I was in charge of (as part of her work on other projects – the authors). Third, the project was not a huge one, and I am good at adapting to new situations, asking the right questions, and making decisions accordingly.

In large projects, responsibilities are usually shared between a project manager and a chief systems engineer. The project manager focuses on managing the client, planning and, control; and the systems engineer mostly deals with the technical issues. In a small project, however, the project manager can also deal with the technical system's engineering aspects, which was what I did."

This was the first time Sharon Shoshany Tavory got to manage a group of people who specialized in different professional areas (the project employed 12 people from

4 different fields): "When it comes to his area of expertise, the specialist knows better than me, but I can ask him which alternatives he had considered; ask him to explain why he had chosen a certain alternative.

This was also the first time I had a budget to manage, and I had to manage the profits in a fixed price undertaking (as a department manager, I had also dealt with finances, but from a planning perspective; the work done was measured by the hour). But I did not work alone; in this area, I used the help of my deputy for planning and control."

In 2003, Sharon Shoshany Tavory was appointed Chief Systems Engineer of the Armaments Systems Division.

Are there major differences between the systems engineering–related functions of a project manager and those of a division's chief systems engineer?

Sharon Shoshany Tavory: "A project is a field activity, where the manager has to choose which systems engineering practices to use in his work (for instance, formulating a systemic architecture, considering alternatives, devising the trial plan and so on). A division's chief systems engineer is different, because it is a headquarters position. Its responsibilities are more oriented towards the discipline than towards its specific applications. Headquarters examines the entire spectrum – all the tools and methods needed by all the projects. This is why a division's chief systems engineer needs to possess a more 'lateral' education and more experience than a project manager. However, it is not always the natural next step in a project manager's career. One can become a chief systems engineer of the engineering division without first serving as a project manager, and vice versa. They are two different positions, with two different areas of responsibility."

The relationship between engineering department and the management of a project is characterized by similar conflicts to those that often arise between the departments in the organization and headquarters units. One such conflict concerns the question of making professional decisions in the engineering unit's area of expertise. On the one hand, the engineering unit's client is the management of the project, which means the unit should serve the project management and meet its needs. On the other hand, the engineering unit is the highest professional authority when it comes to its own area of specialization.

Sharon Shoshany Tavory illustrates this dilemma: "My job as chief systems engineer is not only to provide service, but also to oversee the level of systems engineering implemented in the project. In one particular project, the engineering field specialists recommended a certain mechanical structure material, and the project management refused to follow their recommendation. As a rule, project management has the final word, because it is the client. Traditionally, in Rafael, when the engineering department staff insists on something, claiming to have a well-founded technical recommendation, the project management does its best to overrule it. This is exactly what happened in our case. We appealed, and a team of high-ranking division position holders deliberated the issue and decided to accept the engineering department staff's position."

In 2005, Sharon Shoshany Tavory was appointed the position of Head of Directorate in one of Rafael's product lines. The Head of Directorate supervises the project managers, systems engineers, and design personnel and controls the projects in his jurisdiction.

Sharon Shoshany Tavory: "This position is considered part of the career path of project managers, and not systems engineers, who have no career paths to speak of."

(Note: In our view, systems engineers do have their own career paths, albeit not formal ones, seeing as systems engineering is, as yet, an emerging profession. Systems engineers can be promoted to chief systems engineers or to project management positions that require in-depth knowledge of systems and technology. It should also be noted that although there is a separation between the career progressions of project managers and systems engineers, it is not absolute. In technological organizations that deal with areas like defense, aviation, or space, many of the project managers are former chief systems engineers. This is also the case for Sharon Shoshany Tavory herself. In business organizations, however, project managers are raised from among marketing and business management specialist – The authors).

Sharon Shoshany Tavory: "As a rule, the Head of Directorate only deals with the overarching control aspects of systems engineering, in the projects under his responsibility. Here, too, systems thinking is a useful instrument; but, seeing as systems engineering has always been close to my heart (I suppose the combination of technical and philosophical abstract thinking was what drew me to it), I continued practicing it more deeply than is traditional: in administration projects, at the Israeli Association on Systems Engineering (INCOSE_IL) and in the capability maturity model integration (CMMI) process (a program for assessing an organization's 'maturity' as it pertains to systems engineering and project management) in Rafael.

During my time as Head of Directorate, I learned about systems engineering in product lines, where it includes the various considerations entailed in finding the optimal technical solutions for products that share similar properties.

Explanation: managing the systems engineering of a family of products necessitates the creation of commonalities. This entails the planning of subsystems, able to serve different products that belong to the same family; for example, using the same control system in different vehicles. This is an expression of one of the changes the systems engineering field has undergone. From a discipline that supports one project, it has become a discipline that accounts for the fact that one project begets others. This is a systems engineering that designs subsystems that fit into different products."

We shall present an event from that time that demonstrates Sharon Shoshany Tavory's systemic thought patterns and conduct as a systems engineer who sees a technical problem, gathers the relevant knowledge, and suggests solutions while facing constraints: "One day, during the 2006 Lebanon War, I came to visit one of my colleagues in the missile division, who had been stationed in the control room of the early warning system that warned civilians of imminent attacks by rockets. I was excited, both by the work itself and by the fact that he had such a direct contribution to the war effort. I, too, wanted to contribute, and it turned out they were in need of a quick solution for a connectivity requirement from the Air Force (between the system that detects the launch point and the system that guides the aircraft towards it). He suggested that I sit in on a discussion on the subject the following morning, and I did. The Air Force representatives wanted us to link the rocket detection system to the aircraft, so that they can damage the launchers in real-time, rather than just alert the civilians of imminent impact. Not only that, but they wanted this task accomplished within four days – a very short time.

I volunteered for the task without hesitation, having learned not to 'blink' in cases like these. I received the approval of my superiors, and we got to work. My part consisted of designing the system and coordinating the factors. I was assisted by two software and integration specialists, who were in charge of adjusting the existing system and building the interface between it and the aircraft. Later, another project leader joined the effort. On the fourth day, the system was installed in the Air Force's control center and functioned as required. This task required me to rely heavily on my 'peer leadership' skills, because I had no formal authority over the people I worked with."

In 2008, Sharon Shoshany Tavory had left her position as Rafael's Head of Directorate and became a consultant and a lecturer.

The Relations Between Software Engineering and Systems Engineering

In our retelling of Sharon Shoshany Tavory's career, we mentioned that in the beginning, she and her fellow electronics engineers were asked to develop real-time software, because, at the time, The Department of Digital Systems, to which she belonged, believed they would do this better than software specialists (see preceding text).

Sharon Shoshany Tavory gives an example that explains how this perception came to be: "One of the members of the team I led was an academic software engineer who wanted to work in the field and was given a task: to design the software for a box that contained an electronic system. Every time I visited him I saw a different design. I did not understand this. The previous design seemed fine, so why start over and change everything? I decided to get to the root of the matter and found out that all those times, the integration with the hardware had failed, and he went back to square one and redesigned (and then rewrote) the software. It seemed to me, however, that the problem had nothing to do with the software. When the hardware specialist looked into the matter, he discovered that the system's grounding had been faulty, and so, roughly once a day, when the programmer would run the software he had designed, the system would receive a minor electric shock and the content would be distorted; the programmer interpreted this glitch as a software problem.

Grounding is one of the most common reasons for electronic equipment failures, but only an electronics expert would recognize that he is dealing with faulty grounding. Software specialists, who are strangers to electronics, assume there is nothing wrong with it, conclude that the problem must lie with their software, and decide to rearrange their designs."

But do electronics engineers have sufficient skill in the software field to lead the development of real-time software without seeking help from software experts?

Sharon Shoshany Tavory: "Electronics engineering and software engineering are closely interconnected. Electronics is the base infrastructure of all computers and of the software that runs on them. In the Technion, where I studied, the prevalent approach was that computer engineering stood in the middle between electronics engineering and computer science. 4 out of the 13 programs of study in the electronics engineering faculty were in computing fields.

Of course, one can study electronics and ignore the world of software, but today, most students choose not to skip computer science studies, and do become acquainted with the world of software and its systemic aspects. This is the prevalent approach in all engineering disciplines, today."

Indeed, software engineering has been becoming more and more important in the engineering world. After all, software is a core element of the connectivity between technological systems. As technological ability increases along with the capability of designing complex systems, so does the importance of software (and consequently, of software engineers).

Sharon Shoshany Tavory: "In the past, systems were developed without the help of software specialists; that approach had failed. Today, unlike then, one person cannot grasp and contain the range of engineering disciplines entailed in a complex project, and so varied teams are needed."

She illustrates the increase in the importance of software by giving an example from Rafael's area of activity: "The missiles developed in the past mostly consisted of physical parts, like the warhead and engine, and a little bit of 'brain'. Now, a missile is only one part of a much larger command and control system; the system that controls the missile. Missile development used to be a classic aeronautical project. Today, it is part of a project an aeronautics expert can easily get lost in, even if he does have some knowledge of software.

These transformations have had an impact on the risk management model, as well. In the past, we dealt with such risks as engines that did not allow the required aerodynamic performance; today's risks are systemic. The most problematic area has shifted towards software, because of the increase in the importance of connectivity (which software facilitates), which, in itself, is a source of complications. All these make the system harder to understand for those who specialized in other engineering disciplines.

In the reporting sessions of the past, you could often hear statements like '... let the software people make their report last, it's not really interesting, and we want to get this done with and leave'. This is not an option today, because software is embedded into every field."

This suggests that the bilateral connection between software engineering and systems engineering has really existed all along. Since systems engineering is a methodological tool, meant to support the planning of engineering systems and link them together, a systems engineer has to understand and show interest in the engineering disciplines that make up the system he is in charge of. Software engineers, on the other hand, have often been thought of as specialists, concerned only with software, unfamiliar with the traditional, physical world of engineering.

Systems engineering and software engineering bear a resemblance to each other, seeing as they are both, in essence, abstract systems. Both disciplines need virtual models that allow them to accomplish at least one of their shared overarching missions: to create integration between technological systems.

Sharon Shoshany Tavory: "In systems engineering, there is a lot of abstract discussion that needs to take place before moving on to the 'physical' stages. It is different from classical engineering fields. In many cases, engineers who focus on the physical aspect and attach little importance to the abstract aspect are criticized. Software engineering, however, is abstract in its very essence. Software specialists do not produce a physical product. The blueprint of the product – the code – is the product. And so, in time, the work methods of system engineers have become more and more similar to those of software specialists.

On the other hand, seeing as software engineers do not come into contact with any other technology except software, they are often unable to comprehend it. This is why, back at the department of digital systems, we wanted electronics engineers to be the ones to write real-time software."

Has the similarity between these two engineering discipline brought about mutual adoption of work methods?

Sharon Shoshany Tavory: "Indeed, it has. For example, the method called 'System Modeling Language' or SysML. SysML is a computerized visual modeling tool that allows for easier understanding and testing of ideas, before there is a system to speak of. Originally, the tool was developed to be used by software developers, who refer to it as 'Unified Modeling Language' or UML in short. When we first discovered UML, we saw that it could be used for systems engineering. We used it not only for modeling software, but also for modeling systems. Sometime later, INCOSE used the UML as a foundation for an expanded systems engineering tool, and SysML was born."

Insights on the Systems Engineer Position

- A systems engineer needs to communicate with other disciplines. One can only lead the development of a system when he leaves the confines of single-disciplinary engineering. I did not just perform integration with other engineers; I also worked in their areas, on subjects outside my area of expertise. I did this, not just because it was required in practice, but because I was willing to venture into those areas.
- On the correct mix between the three factors that affect the development of a systems engineer: personality, experience, and training:
  Many systems engineers have a knack for good systems engineering and practice it intuitively. For these people, personality and experience are crucial factors. One of the most talented systems engineers I have met had also been formally trained (in the Technion's ME program in systems engineering), despite the fact that he possessed considerable experience. He said to me: "The course gave me tools to express myself with; I needed that supplementation."
  The order of things is also important. Systems engineering training should only be had after gaining hands-on experience, otherwise it offers no benefit. A systems engineer first needs to practice integration processes, see systems fail, and understand why it happens. If he does not encounter these situations, he will not be able to absorb the lessons of the training program.

To conclude, a systems engineer's personality, his ability to see things from a broad perspective, is very important. Having that, he needs to gain experience in order to become more aware of potential problems. Finally, formal training is the factor that completes a systems engineer's expertise. Personally, I found the fact that I majored in technological subjects in high school helpful, because there, I was introduced to various fields and gained some hands-on experience as well.

- In some cases, people want to become systems engineers because they think of it as a promotion. But, compared to project management, which certainly is a promotion, the status of a systems engineer depends on the discipline's positioning within the organization. Companies do not always know how to position the job, and systems engineers often have no options for promotion beyond their current position. (See aforementioned author's note on this subject).
- There is a list of generally acceptable qualities a good systems engineer should possess. Not all of them are needed to succeed on the job, but effective communication is definitely among those that are.

3.4

SYSTEMS ENGINEERING AND ACADEMIA

3.4.1 "APPLYING HOLISTIC THINKING"

An interview with Prof. Joseph Kasser

As a field in its infancy that has yet to complete its evolution and is in the process of defining itself, it is only natural for systems engineering to be perceived differently by different people. Even now, some see systems engineering as a profession, at the base of which lies the development and implementation of engineering processes, and thus, only those who have training as engineers can practice it. The other end of the spectrum is occupied by those who do not see systems engineering as a profession at all, but rather as a collection of tools that helps the formulation of thought processes for more efficient problem solving (problems which are not necessarily related to engineering). As such, systems engineering can potentially help anyone, not just engineers.

One of the loudest voices siding with this approach is that of Prof. Joseph Kasser. An electronics engineer by training, he turned to systems engineering later in his career. Today, he is a prominent researcher and lecturer on systematic and holistic thinking.

We have conversed with him about his perception of the field.

Managing and Engineering Complex Technological Systems, First Edition.
Avigdor Zonnenshain and Shuki Stauber.

The General Approach

Prof. Joe Kasser's words suggest that his perception is comprised of three concentric circles: the largest circle is *holistic thinking*. It is a way of thinking that sees the system from various perspectives: not just from within, but from without as well. This approach allows one to see the system from unusual perspectives. This way, things that, at first glance, may appear as difficult problems become the foundations for creative solutions. Holistic thinking can serve anyone who has a problem to solve.

The second circle is *systematic thinking*, which is, in fact, part of holistic thinking. It examines the system, its components, and the relations between them. It also examines the processes at work within the system. Systematic thinking is relevant for all systems, not just for technological, engineering-oriented ones. For instance, an organizational unit is a system too, and managing employees requires systematic thinking.

The third circle is *systems engineering*. Systems engineering is the application of holistic thinking, often, but not always, implemented in systems that are technological at their core. This is why there are many who claim that a systems engineer must have training in engineering.

Background

Systematic thinking has accompanied Prof. Kasser from days as a teenager: "Even back then, I understood that I saw things differently than others. My father, and my teacher, taught me to see things from other people's points of view."

All his life, he acted as a holistic thinker, even as an organizational specialist, whose job description did not include the word "system".

Joe Kasser: "During the 70s, I worked as a systems engineer for Bendix Aerospace – then, a subcontractor for NASA, engaged in building the science package the astronauts took to the moon on the Apollo 15, 16, and 17 missions. I worked in accordance with traditional systems engineering work patterns, and I still managed to find creative solutions for the problems I faced."

During the 80s, Prof. Kasser worked in Israel for Luz Industries, a company that developed and established solar fields. He joined the company as head of electronic control and monitoring and was also the lead systems engineer of the solar field.

He says: "All the odds said Luz was supposed to fail as a startup, because of the combination of distance, communication, and language constraints. But it succeeded anyway. It took me 15 years to figure out why – because we used holistic thinking in our systems engineering. That is what made the difference between success and failure."

Prof. Kasser demonstrates the use of holistic thinking in a company: "We needed to control a field of mirrors using computers. The computing costs were supposed to reach nearly a million dollars, a very large expense. We looked for a solution and, by thinking outside the box, found a way to use concepts from satellite communications. We saw the similarities between the satellite network and the mirror controller network.

This is an example of looking at things from a different perspective. Not only did we lower the computing costs to $2,000 per control system from $300,000, we were also able to simplify the software."

His last job as a systems engineer was at Ford Aerospace. This was when he began his doctorate studies, as part of the company's employee training program. In 1997, Prof. Kasser founded his own consultant firm, graduated, and became an academic researcher and lecturer.

Prof. Kasser does not only lecture and write about looking at systems differently to find creative solutions; it is also evident that he uses this approach in his work as a researcher and lecturer.

Thus, for instance, his dissertation, done in the engineering faculty, examined systems engineering from various angles, not just those that pertain to engineering: "it was a combination of systems engineering and business administration. It was based on the assessment that a group of small businesses, each responsible for different tasks, can, in effect, function like a large company."

He adds and demonstrates: "When I began reading systems engineering textbooks, I discovered that they were full of contradictions. I particularly remember a symposium, where a number of lecturers had presented their perception of the field. Each of them had defined systems engineering differently.

What stood out particularly was that I could not find anything that systems engineering did that was unique to systems engineering. Everything that was defined as part of systems engineering was also done by someone else: financing, testing, design, architecture ... all those were performed by other people in other fields. A year later, I published an article entitled 'Systems Engineering – Myth or Reality'."

Today, he has just completed writing a book that combines systematic thinking and holistic thinking, entitled "Holistic Thinking Creating – Innovative Solutions to Complex Problems."

In the second half of this chapter, we will present Prof. Joe Kasser's perceptions on systems engineering in its wider context.

Beyond Systems Engineering

On holistic thinking: "Most people examine something from a single perspective. If they like it, they adopt it. Their minds don't examine any further aspects, they don't look around. I call it 'the perspective parameter' – how your mind looks at the problem from different perspectives. A problem needs to be examined from all sides. If we do that, we will see more solutions than problems.

Holistic thinking allows people to see solutions where others see only problems. Holistic thinking is important for understanding things related to creative solutions.

Holistic thinking deals with more than the internal and external perspectives and how they relate to one another. It examines the way systems work, how they are used, their liability and more.

In order to get a good system and find solutions to problems, it is necessary to go to the roots of the system and compare it to other systems. This way, it is possible to discern common qualities, positive and negative ones, to see the differences and figure out the reasons for them.

It's not enough to look at things from the inside, at the structure and functionality of a system. To find solutions, one needs to look at things from more perspectives.

Holistic thinking for problem solving is suitable for any situation, from control systems development to tracking a person on the moon to properly holding a glass of water – if you put your pinky underneath the glass of water you drink from, it's because you understand that the older a person gets, the less steady his hands become. By placing the pinky underneath the glass, you minimize the chance of the glass falling. This is an example of risk management viewed systematically.

Anyone who wishes to solve a problem needs to think holistically. The way to create the solution is systems engineering, an application of holistic thinking."

On systematic thinking: "Thinking systematically means looking at the different parts of a system and how they relate to one another.

Systematic thinking means two things: the first, thinking about the system and examining the relations between its components. The second, thinking systematically.

The question is: where do we start defining the system? Defining the boundaries of the system is critical. For one person, the system is the car; for another, it is the car and its passengers; for a third person, all the cars on the road are the system. The engine is also a system. This is where systematic thinking is needed."

On Systems Engineering

On the essence of systems engineering: "Systems engineering is a way of thinking for solving problems. It includes a set of tools, solutions and alternatives, processes and people. This problem solving process is the process of systems engineering.

It is important to make the distinction between the role of systems engineering or the job title and the activity of systems engineering.

Systems engineering is, in essence, the application of holistic thinking. It's called systems engineering because we tend to use the wrong words. People come with a concept and don't know there already is a word for it, so they invent a new one. The word engineering means making something happen. Some see systems engineering as processes. Some see it as a way of solving problems. Others see it as a discipline full of procedures.

Systems engineering is both a profession and a discipline.

Systems engineering is a way of thinking, and so, it is unaffected by cultural differences. Cultural differences only affect how things are expressed. In different cultures, systems engineering is expressed in different ways."

On qualities and abilities: "Good systems engineers see connections that other people miss. This is why they also see solutions.

Every engineer has to be a systems engineer some of the time. He or she won't be using this skill all the time, but when he has to solve a problem, he will need it. Once a solution is found, the rest is engineering.

The good people in the field of systems engineering hail from the world of physics or electronics, because they understand the real world. It is an important component in their training. It is important for a systems engineer to possess training as an engineer, to understand exact sciences. He needs to be one who understands that things cannot be changed simply because we want to change them. Engineers use systems engineering on the technological level, in the construction of airplanes, ships or tanks.

Systems engineers are engineers, because that is where it helps them solve problems. As a rule, one does not have to be an engineer in order to be a systems engineer. One has to be an engineer, if one works in an engineering environment. If a systems engineer practices aerospace systems engineering, he has to be familiar with aerospace. But if the systems engineering pertains to marketing, then he needs to be well versed in marketing, and he doesn't have to be an engineer. Systems engineering is the application of holistic thinking, you can be a marketing systems engineer, if you implement systematic thinking.

A systems engineer needs to know the industry he works with, otherwise he will reach the wrong conclusions and his decisions do senseless things. For example, if he knows nothing about TVs and sees smoke rising from a broken TV, he might assume that TVs are normally filled with smoke and that, therefore, has stopped working because the smoke has leaked out, so the smoke needs to be put back in for it to work again.

A project leader does not have to have systems engineering background. What's important is for him to be able to think holistically."

On the effective application of systems engineering: "People don't usually stop to think what a systems engineer needs to do. If we look only at what they do, without knowing what they need to do, we will never know if we've missed something important.

In effect, today most systems engineers follow processes. Only a small minority are problem solvers."

On Systems Engineering and Management

Joe Kasser: "Systems engineering is a problem solving mechanism that includes many managerial elements, because it involves human components and processes. It is a mechanism adopted mostly by engineers, because they found it useful for solving systemic technological problems.

There is an overlap between systems engineering and management, because systems engineering designs the processes that managers later supervise. There are a lot of professionals who use systems engineering tools – detectives, for instance. Also, cooks, who have to cook a meal where all the ingredients are combined in the right order, the right way, and at the right time.

Managers don't normally use systems engineering tools, because they are not taught to do so. Systems engineering methodologies should be taught in administration schools, not just engineering schools.

Systems engineering is the management tool of the 21st century. It is a different management method that includes tools and techniques suited for each case."

3.4.2 "A POWERFUL NATURAL CURIOSITY AND AN ABILITY TO TRULY LIKE PEOPLE"

An interview with Dr. Cecilia Haskins

Despite the fact that systems engineering is a young discipline, a sizeable number of people already hold job titles as "systems engineers" and many higher education facilities now offer programs that grant advanced degrees in this field.

Dr. Cecilia Haskins, an American teaching systems engineering in Norway, is among those who believe that a systems engineer does not have to be an engineer, and the appearance of the word "engineering" in the name might in itself be a mistake. The really important word is "systems" – relating to the ability to see the big picture and act systematically. In this chapter, Cecilia Haskins describes her perception of systems engineering.

Who is a Systems Engineer?

Cecilia Haskins has always been a systems engineer: "I was born a systems engineer. From a very young age, I have been very systematic in the way I approached everything. I had the ability to organize other people's work as well as my own. When my father, who also had a systematic mind, would return from work and talk about his day, I would listen and ask questions."

At the young age of 15, Cecilia already knew she wanted to pursue a career in Electronic Data Processing (EDP), but at the time, the United States did not yet have a university that offered education in this new field: "And then I got some great advice from my father – advice I now pass on to others – to study natural sciences, especially chemistry, the perfect background for systems engineers. The basic principle of chemistry is that atomic elements are very small, and everything else is made from these. This is why it is the perfect mental model for systems engineering. In my 40 years of work, systems engineers who started out as chemists were the ones I enjoyed working with the most."

It is evident from Cecilia's words that the ability to be a successful systems engineer is, first and foremost, inborn. Cecilia specifies further: "A systems engineer needs to have two key traits and both are critical for good performance: the first is *natural curiosity*, namely, never getting tired of asking questions; the second is truly and deeply *liking people* – no matter who they are or what your impression of them is – having the desire to do things with them, and learn from them."

She gives an example from her work as a student advisor: "Today I am involved in a project, where graduate students manufacture a car. It is a long process, currently in its third year. Three years ago they realized that a project of this scale needed to employ systems engineering, and I was charged with teaching them these skills. The first student who volunteered for the project was a born systems engineer. Nobody even came near his level of ability. I hardly had to advise him at all. He instinctively knew what to do to help his team succeed. All I had to do was to refer him to the literature so that he could write his dissertation. He did things naturally, and when, later, he was exposed to the literature he saw how the things he had done were formally defined. The other students who came after him were different. They studied first and then said: 'oh, we should do something about those requirements.' They used the literature, what we had taught them, and they did it very well, but it didn't come naturally to them like it had for the first student."

Another "amusing" (in her words) example of the two key traits required of a systems engineer is taken from the beginning stages of Cecilia Haskins' career: "I started my career with a summer job as a programmer analyst in the early 70s. In my first year, my boss asked me to develop a program that included certain components. As usual, I started to ask him a million questions he was completely unable to answer. And then he said to me: 'if you want answers to all these questions, go ask the client yourself.' Thus, at the age of 17, my curiosity and interpersonal skills had qualified me to sit with a client with no one else present. These two central traits are essential to be a systems engineer."

There are many reasons Cecilia believes that a systems engineer does not have to be an engineer: "We have many systems on our planet that do not require knowledge of advanced calculus. Moreover, stereotypical engineers love their work but have minimal inter-personal skills. Often, people are offered the opportunity to become systems engineers and fail because they lack natural curiosity (beyond the specific task they are charged with) or because working with other people is not high on their priority list and conflicts arise."

The Essence of Systems Engineering

Cecilia Haskins believes systems engineering to be a combination of discipline, worldview, and profession that offers various ways of problem solving: "Systems engineering offers a structured discovery process for solving problems. The goal of systems engineering is to always see the big picture and do what needs to be done systematically, so as not to miss anything important.

Conway's Law suggests that the way we organize our company is the way we intend to solve our problems. This is what makes the company successful in the long term. However, organizations today tend to be hybrids: they have a hierarchy, but at the same time, they empower their employees. Over the years, people have learned that if they do things in certain ways, they get better results. Systems engineering thrives as long as people are convinced that it is important to understand, validate, and implement requirements, and to think about solutions in terms of architecture."

Cecilia Haskins created a simplified five-stage systems engineering model as part of her PhD research. This would not be sufficient for complex or critical systems, but is sufficient for many everyday problems.

The five stages of the model are as follows:

S – Stakeholders: Identifying the people who are impacted by or can impact the problem;

P – Problem Formulation: Making sure you are treating the problem, not a symptom;

A – Analysis: Defining solution options, creating alternatives etc.;

D – Decision making: Deciding which of the alternatives to employ and in what order (what to do first, what to do later);

E – Evaluation: Continuously evaluating what we think we know.

On Systems Engineering and Project Management

Cecilia Haskins believes systems engineering has a symbiotic relationship with project management: "Systems engineering and project management are united from the start. One of the major problems for managers managing a project is that you don't always see the big picture. You spend a lot of time dealing with budgets, schedules, and deadlines. I have observed project managers take nonoptimal decisions just so they can say 'mission accomplished.' Systems engineering means helping projects succeed. To me this word is key – succeed; to make a project successful.

In the past, no separation existed between these two fields. The managerial and technological components were handled together intuitively. But today, we get to a level of specialization so high that everyone is immersed in their own field and people become disjointed. Project management and systems engineering are like yin and yang (complementary opposites – a term taken from ancient Chinese philosophy – the authors) – one cannot succeed without the other."

The Development of Systems Engineering in Norway

Cecilia Haskins, who lives and works in Norway, tells of the development of systems engineering in that country: "In Norway, the term was introduced in the mid-90s, but systems engineer job positions have only begun to emerge in recent years. Slowly, companies are beginning to understand that this tool is vital for projects to succeed, for good products to be manufactured, for successful systems to be created. But these are early days; there is still a long way to go. Managers are also beginning to understand the important symbiosis between project management and systems engineering."

The oil and gas industry is the most prominent in Norway and the first to recognize the importance of systems engineering after the Norwegian defense.

Cecilia Haskins: "The oil and gas industry is already proficient in many engineering processes related to extraction and construction. In time, they realized that the construction industry was not organized well enough, and often inefficient. For products located in extreme environmental conditions there are many challenges, both technological and physical. They have begun to recognize the fact that systems engineering can help find solutions to some of those problems."

For example: "I had a graduate student who was investigating a company performing a project for the oil and gas industry. The company worked in two organizational groups. One group would talk to the customers before the product was delivered, at the design phase; while the other handled the product's useful life period of 25 years.

The surprising discovery was that these two groups rarely interacted or made use of formal channels for sharing information. My student asked them whether they had thought to look at their product in terms of a system life cycle, so that in future projects, each group could learn from the other group's experience. They said 'great idea!' and accepted his suggestion."

The willingness to embrace systems engineering stems, in Haskins' opinion, among other things, from the change of generations of executives in the industry: "Today there is a senior management layer in the process of retirement. These are people who come from the traditional backgrounds, very different from the background of those arriving to replace them. In addition, engineering schools today pay a lot more attention to the interdisciplinary approach, even if they don't call it systems engineering. Students today know that talking to the other engineers working on a project may be the right thing to do. The new generation is more open to systems engineering, it makes more sense to them.

There is a company in Norway that has recruited many of my former students. The great thing about this company is that they did it slowly, gradually. For instance, former students of mine had to work in three different positions before joining the corps of systems engineers. The company understood that students straight from college do not start out as brilliant systems engineers. It was a number of years before they put the words 'systems engineer' in a job title, and that is how it should be."

3.4.3 "EXPANDING THE BOUNDARIES OF THE SYSTEM"

An interview with Prof. Olivier de Weck

After roughly 10 years of education and training, during which he worked as an engineer and a systems engineer in the aviation industry, while completing his bachelor's degree studies in engineering and his master's degree studies in systems engineering, Olivier de Weck became a prominent researcher and lecturer in the systems engineering field.

We spoke with him about the nature of systems, the connection between systems engineering and the academy, and the developments of the systems engineering profession, as it is integrated into various industries.

Personal Background

The environment Olivier de Weck had grown up in provided a fertile ground for his development as a systems engineer. He was born and raised in Fribourg, Switzerland; a city where half the citizens are French-speaking and the other half German-speaking. But his mother, a translator, and his father, an immunologist, spoke English to him at home.

Olivier de Weck: "Even as a child, my environment was multilingual. Nothing was uniform. I had connections with different countries, different occupations, different cultures, and different languages. The environment I had grown up in was fertile ground for learning about life in a culturally and linguistically complex environment."

But it was not just the environment, but his heart that had led him to systems engineering. Attracted to the aviation world from a very young age, he joined the air force after being drafted into the military (Switzerland has a mandatory militia conscription system similar to Israel) and, after a training period, became an airplane technician.

At first, he was responsible for the F-5, a two-engine plane, considered relatively "simple" to maintain.

Olivier de Weck: "My passion for systems engineering emerged when I started troubleshooting. Pilots would return from a flight and complain of abnormal phenomena, like vibrations during the flight or low oil pressure. They would fill out a flight report with a description of the malfunction, which would then be passed on to our crew, and it was up to us to solve the problem. Today, planes are a lot more complex than that, with an automatic error code displayed for each malfunction. Today they tell you: 'code 538' and you know exactly what the problem is. But back then, nearly 30 years ago, I had to dig through the books and try to figure out the source of the problem. We knew the symptom was the problem, but we did not know the reason behind it, so we had to troubleshoot.

I didn't specialize in a specific area. I had to know enough about all the plane's subsystems to point out the source of any problem. It's a lot like solving a crime: you raise a hypothesis and then you put it to the test. It was an intellectual challenge. I had to absorb a lot of information very quickly. I loved it."

As part of his military service, Olivier studied mechanical engineering at the Swiss Federal Institute of Technology (ETH) in Zurich. He was discharged in the late 90s as an officer, in charge of the maintenance of an entire squadron.

Olivier de Weck agrees that that was the point when he began adopting systems engineering work patterns, but it was only toward the end of his engineering studies that he discovered the discipline, after reading a book written about several authors, based on a fundamental systems engineering book written by Ralph Hall in 1962.[1]

Having completed his military service, he moved to the United States, where he spent four years working at the aircraft manufacturer McDonnell-Douglas in St. Louis. His last position was engineering program manager for the Swiss FA-18 aircraft. He received his master's degree in systems engineering from MIT, where he proceeded to more advanced studies and became an associate professor, researcher, and lecturer.

Olivier de Weck sees himself as both a practicing systems engineer and a researcher of the field. According to him, despite practicing systems engineering in effect, he is able to "zoom out" and see the bigger picture of the system engineering

[1] Reference to the German textbook on Systems Engineering: http://www.amazon.de/Systems-Engineering-Grundlagen-Reinhard-Haberfellner/dp/328004068X/ref=sr_1_1?s=books&ie=UTF8&qid=1366637658&sr=1-1&keywords=Haberfellner.

field. He believes those who have never practiced systems engineering cannot always research it.

On the Nature of Systems

The system's boundaries:

One of the central dilemmas for anyone who practices systems engineering, including the systems engineer himself, is that of the system's boundaries. Thus, for instance, one can decide that the boundaries of the system are the client's technological requirements for the project. Or, one can define the boundaries to include the technological system's impact on the environment. Of course, such decisions can radically change the system's characteristics and design.

Olivier de Weck says that: "Traditional systems engineering had always been inward focused. It made sure all the system's components (the components, the processes, the subsystems) worked together to produce the system's end products and satisfy the requirements set forth by the client. Systems engineering never gave much importance to what lay outside the system.

The problem is that many systems become composed of other systems (see interview with Hillary Sillitto for a discussion on this subject). They become very large and more and more complex. We start building systems like we have never designed before; connecting systems that were not designed to work together. We do this because we think to derive some benefit from it."

He demonstrates: "Drivers texting behind the wheel is the most common reason for traffic accidents in the United States (in the past that reason used to be drunk drivers). This has to do with systems engineering, because if you analyze the problem, the traditional transportation system is now joined with communication systems in unexpected ways and means of communication, with human behavior and motivation at the center of it all."

How the boundaries of a system are defined affects the design of the solution to the problem: "If we look at a system as purely technological, we may be able to limit the device's communication abilities when it is in motion. But if we expand the definition to include the entire transportation system, the solution can also be preventive legislation."

All this, de Weck sums: "has to do with systems engineering because when there is a problem with the system, a system-wide solution is needed." (Of course, technology is at the core of the combined transportation system, so the issue still lies in the systems engineer's playground – the authors).

Expanding the boundaries of a system also allows us to see connections we cannot always see from within the system itself. Olivier de Weck: "We design a system with certain boundaries and see no correlation between A and B, but if we expand the boundaries, we can suddenly see a correlation (or synchronization) outside the original system. Seeing this correlation has a profound effect on how the inside of the system is designed. For this, the concept must be modeled in greater detail."

Between short-lived and long-lived systems:

The differences between systems with a limited lifespan (such as development projects) and those designed to operate over long periods of time, with no expiration date (such as toll roads or the Internet) have already been addressed elsewhere in this book.

While "short-lived" systems, according to Olivier de Weck, are artificial frameworks designed to achieve concrete objectives, "long-lived" systems can evolve like living organisms and are more difficult to control.

He says: "these systems are partly planned, partly evolved. Some parts of them are designed as artificial systems by systems engineers, while other parts need to be addressed in a way that simulates biological thinking, with the concept of evolution in mind. Evolution cannot really be controlled, but one can plant good genes in the system."

To the question whether the maintenance of a "long-lived" system (maintenance is usually less complex than development – the authors) requires the use of system engineering tools, Olivier de Weck says that it depends on the complexity of the system. According to him, a simple system, like a road, does not require systems engineering (from here we can conclude that a toll road, which includes collection systems, electronic monitoring systems, rescue crews etc., may indeed require the use of such tools – the authors), but a more complex system, like a refinery, does need them.

He agrees with the statement that in certain cases, a system's maintenance can be seen as an ongoing project that requires constant troubleshooting: "If we want the system to continue operating at a certain level, we need to upgrade as well. And then it becomes a sort of gradual development process."

Systems Engineering and the Academy

When discussing the question whether systems engineering is just a collection of disciplines, a job, or a profession, Olivier says that systems engineering is indeed a profession, albeit a young one (we called it "profession coming into being" – the authors), that has only existed for about 20 years.

He describes the evolution of the profession within the academic framework: "Traditional systems engineering is processes oriented. 'Follow these steps and these instructions and meet the standards.' Before World War II, engineering had been a list of specified actions, of empirical rules of thumb following an empirical recipe. During and after the war, larger, more complex projects, like the Manhattan project, began to emerge, in which most of the leaders were not engineers, but physicists. At the time, a very influential article had been published, which stated that systems engineering could no longer base itself on empirical rules of thumb, and had to rely on the laws of physics. The academy was forced to reexamine itself. Engineers began switching to a more analytical direction, aiming to understand the physical and chemical bases. Suddenly, the difference between what had been taught to chemistry students and chemical engineering students was not so great anymore. Engineering moved toward science, more analytic than synthetic (on this subject, see interview with Prof. Aviv Rosen).

Only in the last 20 years, with the birth of systems engineering as a research field, has the academy begun to understand that synthesis is also necessary in development. When I discovered systems engineering, it was already about more than just problem solving, but implementing systematic thinking in the technological world. The direction it was heading towards was how to pick an amorphous situation, model it at the right level of abstraction and carefully dissect it into its smallest components that could be solved, then reintegrate into a working solution."

On the same subject, he had this to add: "Systems engineering is a new discipline based initially on trial and error, empiricism. There was a study that suggested that if less than 12% of a project's budget is invested in systems engineering, the project is bound to encounter problems, because important things will be missed. In this situation, the project leaders are sure to face unpleasant surprises. In contrast, investing more than 18% of the budget in systems engineering would cause the project to become too cumbersome, making systems engineering a burden and too bureaucratic. These findings are also based on empirical observations, on experience – these are not proven theories yet, they have no scientific basis.

There is a research field called 'systems science.' Many of those who practice it are experimental physicists and mathematicians. They model systems in a very abstract fashion. Often systems are represented as binary networks using graph theory. Those who work on projects in the field don't know what to do with that. They say: 'this is interesting, but it does nothing to help me'."

Olivier de Weck claims that there is a significant gap between the theoretical models of a system and practical systems engineering, which relies on processes and hands-on experience. This gap will take some time to bridge, but progress is being made in this direction.

The Human Aspect of Systems Engineering

Olivier de Weck is of the opinion that the human factor in systems engineering is underestimated. He agrees with Technion Prof. Aviv Rosen's claim that one of the reasons for this is the fact that it cannot easily be described in mathematical terms (an environment engineers find convenient to function in) and is difficult to oversee and control.

He says that one of the most important issues in systems engineering is the decision concerning a system's level of autonomy, namely, the act of differentiating between actions that should be performed by man and those that should be performed by machine.

Olivier de Weck: "For this purpose, it is important to understand when people function at their best. There is a curve that nicely describes the relationship between people's effectiveness and how busy they are. The optimal level is achieved at around 85% cognitive loading. This leaves people with another 15% of their time to handle unexpected things and emergencies. Beyond this level of cognitive loading, people become overloaded and make mistakes out of pressure. In addition, such an environment causes people to become unhappy and worn out.

On the other side of the spectrum, some studies suggest that when automation levels are very high, people have very little to do and get bored. Examples are operators of power stations. They become distracted and fail to pay attention when needed, which decreases their effectiveness. For this reason, when systems engineers design a system, they must place a heavier emphasis on human limitations. We have only recently begun to realize that the right way to design systems may revolve around human ability."

He gives an example of a system that failed, because it did not consider the complexity of human behavior: "To resolve its air pollution problem, Mexico City decided on a new policy that would decrease the traffic in the city by half. They designed a new transportation directive that directed anyone with an odd license plate number to only be able to drive on odd days, while those with even numbers could only drive on even days. The result was unexpected: not only did the air pollution levels not decrease, they rose. The reason was that a large number of people simply purchased a second car with a license plate that allowed them to drive during the remaining days of the week. To minimize the extra expenses, people tended to buy old cars, which tend to pollute more. The legislation did not take into account people's ability to circumvent it and was eventually abolished."

Systems Engineering in Different Industries

It is apparent that systems engineering mostly aims to find solutions for complex systems, those that need organized work processes so that possible malfunctions can be prevented and handled successfully when they do occur. From here, the differences in the use of systems engineering in different industries are derived.

Olivier de Weck finds that systems engineering is especially developed in the aeronautics industry, as well as in the submarine and nuclear reactor industries, because these are industries that require very efficient or very safe engineering.

Olivier de Weck: "In these systems, you cannot rely on what you see and say 'it looks OK to me,' you have to be very accurate. One such industry that failed to adopt this approach properly is the oil and gas industry. A lot of offshore oil drilling takes place in shallows, but major incidents, like the BP Oil Spill (British Petroleum, one of the world's largest energy companies – the authors) in the Gulf of Mexico, happen in deep waters. These drilling projects are complex systems that have to handle extreme conditions, not unlike those of space exploration, namely, working with robots under high pressure, at high temperatures, and at distant locations. In spite of all that, when asked about systems engineering, the people of this industry usually respond by asking what that is. This needs to change.

The first signs of the implementation of systems engineering are beginning to emerge in refineries founded today, but things are still done sloppily, and the dangers are many. When the system operates at low temperatures and pressure levels and there is a leak, the leak is repaired and the problem is resolved. But when the pressure and temperatures are high, the same leak becomes a serious problem. This industry has only now begun to understand that it cannot go on this way." (see discussion about this industry in the interview with Cecilia Haskins).

Olivier de Weck sees a significant difference between the business sector and the public sector, in terms of the sector's willingness to adopt systems engineering work patterns.

He says: "Systems engineering in the public sector, in government or defense projects (which usually are also government projects – the authors), is integrated into the system, an inseparable part of the requirements specification. The business sector, on the other hand, is focused on immediate or short-term benefits, and so, only uses systems engineering methodologies if it has added value, namely, financial profitability.

The problem with systems engineering in the business world is that its short-term benefits are somewhat hidden. It is mainly viewed as an extra business expense. Even if great efforts were invested into systems engineering, the benefits will only emerge after a period of time which could be several months or years. When a complex system lasts many years, people will talk about what an impressive job the systems engineers had done on it, and how they should be thanked and appreciated for it. But after so many years, those systems engineers will not receive the recognition they deserve, because by then they will have retired or moved away. The gap between cause and effect here is very wide. There is a need for more research on quantifying the value of systems engineering."

In this regard, Olivier de Weck agrees that systems engineering is akin to preventive medicine.

More Insights On Systems Engineering

On the qualities of a systems engineer

- Olivier de Weck's words suggest that not anyone can be an effective systems engineer. Alongside his understanding of the technological, systematic aspect, a systems engineer must also have very good interpersonal skills. He must understand the fields he is engaged in and, at the same time, be open to collaboration, because "by their very nature, systems engineers intervene in other people's business. They look under your carpet, to see what you swept under it. This is why systems engineers must be not only knowledgeable, but also possess the right character for the job. Systems engineers who are less knowledgeable and more bureaucratic, those who strictly follow procedures, are not always popular people. Such people are seen as a nuisance and a bureaucratic overhead."
- In de Weck's opinion, a systems engineer has to have the basic education and training of an engineer, because he needs to be trained in planning, designing, and operating technological systems. Oftentimes, he or she has a home discipline like controls, structures and materials, or software engineering, to name a few.Good home disciplines for systems engineers are those that tend to touch all or most of a system. This is true in cases when technology lies at the core of a system – be it an engine or a database.

On the essence of the job

- One of the major questions concerning the definition of the profession is whether a systems engineer is a specialist or a generalist. Olivier de Weck believes that the most common assumption is that a systems engineer is a generalist, but a fundamental problem is apparent: a systems engineer does not start his career as one, but rather becomes a systems engineer after gaining experience as an engineer. "Companies recruit systems engineers as engineers first; then, as part of the organization's career paths and rotation, they go through a number of positions and eventually become systems engineers, having learned a little bit of everything. In my opinion, that is not the best way. Systems engineering is and should be a specialization in itself, like software engineering or thermodynamics. Systems Engineers are specialists in formulating proper requirements, deriving verification plans, managing complexity and interfaces etc ... this is not a general skill, but an abstract set of skills that require specialists."

On a systems engineer's professional background

- Another important point already raised in other chapters of this book (see, e.g., the interview with Mimi Timnat) is whether a lead systems engineer in a project needs to be an engineer hailing from the project's core discipline. For instance, should a project that is, at its core, mechanical be led by a systems engineer with background in mechanical engineering?
  Olivier de Weck: "You can't be a good systems engineer if you only know a little about each discipline utilized in your project. There are fields, in which one must be more knowledgeable, while in other fields, one can get by with less knowledge. It is important to be well versed in at least some fields; otherwise it is impossible to see nuances that are vital for understanding the system. A lead systems engineer who only knows a little about each field will be perceived as overly naïve.
  If there is a project, the professional core of which is outside the areas the lead systems engineer is well versed in, but does relate to them, then that seems alright to me. This has to come alongside a deep understanding of the fundamental processes of systems engineering, such as requirement definition, interface management, system examination, and system assessment. These are all areas in which the systems engineer must have in-depth knowledge and understanding."
- For this reason, Olivier de Weck believes a systems engineer cannot easily move from one industry to another. "The knowledge a systems engineer accumulates in a certain industry may not be as relevant in others. Switching between industries like that is what creates the situations where an engineer's knowledge in most of the fields the new industry's project concerns itself with is minimal, and he has to start educating himself on a variety of subjects."
  (On this subject, see discussion with Mimi Timnat)
  "However", he adds, "systems engineers are quick learners and many approaches and methodologies can be applied to more than one industry. Of course, not all industries adopt them, and when they do, they often do it differently."

(Among other reasons, this is because there are no uniform rules yet for working with the various engineering fields, and systems engineering standards are only now being formulated – the authors).

3.4.4 "A PROFESSION MEANT TO SERVE THE NEEDS OF THE INDUSTRY"

An interview with Prof. Aviv Rosen

Prof. Aviv Rosen from the Faculty of Aerospace Engineering at the Israel Institute of Technology (the Technion) is also the head of The Gordon Center for Systems Engineering, home of the Technion's systems engineering graduate program. In our interview with him, we discussed his perception of the discipline's evolution, its character, and its importance; the connection between aeronautics and systems engineering; and the training of systems engineers.

The Two Overarching Domains of the Engineering World

Prof. Aviv Rosen finds that, by way of generalization, the world of engineering can be divided into two domains – analysis and synthesis: "Analysis is associated with the world of research, and its products are usually models for understanding various phenomena. Innovations often begin with analysis. Conversely, synthesis is the ability to bring components together and produce an engineering product. This is usually done by the industry. Synthesis is considered to be of a more routine nature, and was therefore perceived as inferior to analysis by academia for many years. Research was thought of as a more lucrative practice, as it offered the possibility of discovering new things and publishing one's findings in scientific magazines.

But times have changed, and the importance of synthesis has slowly increased in many engineering fields. Consequently, the rate of appearance of major technological innovations in these fields has diminished. Take, for example, the jet engine: it conquered the market after the end of World War II, and still remains the most common means of airborne propulsion, sixty years later. Today, there are a lot of similarities between different types of aircraft, and success is no longer determined by the level of technology (as it does not vary much between the various manufacturers), but by how to better integrate the system as a whole, that is – by synthesis.

While, in the past, the market wanted the best technology and paid little mind to the cost, today the orientation is also – and in many cases, mainly – economical. Technologically, the industry has everything it needs. As far as it is concerned, today's added value is in the benefits of higher profitability. Competition is no longer over breakthroughs, but over who has the best systems engineering."

Systems Engineering is More Than Just Engineering

Prof. Rosen's first direct, unmediated encounter with systems engineering happened when, as the Dean of the Technion's Aerospace Engineering Faculty, he was presented with the industry heads' request to found a training program for systems engineers (more details on this next). But, according to him, the turning point in his personal views on the discipline was when he first began to understand the importance of the 'soft sciences': "I come from the world of exact sciences – mathematics, formulas, and computer programs – and I have learned that, in many cases, the so-called 'soft' sciences are no less important than the technology. If the technology is superb, but the economics are poor, the product will not sell. If you have created a technological wonder that does not suit the market, you have failed. At the end of the road, there is always a client, and his psychology needs to be taken into account; products must sell.

Systems engineering knows how to take these things into account and integrate them into the engineering process. The engineering and the needs of the client have to go together, and this combination is what systems engineering is about. For example, one of the discipline's most salient terms is 'requirements management.' In the past, engineers did not give requirements the weight they deserved; they just developed as well as their professional abilities allowed them to. Today, it is clear that system requirements cannot be ignored, and must be a pivotal factor throughout the product's life cycle. This dramatic change in perception was instilled by systems engineering, and the new approach has now become part of the industry's methodology."

Systems Engineering and its Affinity with Aeronautics

Other chapters of our book have addressed the claim that systems engineering has evolved hand in hand with aerospace engineering. Prof. Aviv Rosen expands on this: "An aeronautical engineer's training is very similar to that of a systems engineer. Even during the 60s, back when I was a student in the Technion's aerospace engineering faculty, there was a conflict between the need to specialize in one of the various subdisciplines of aeronautics, and the need to train a 'well balanced engineer,' with background knowledge in all areas. The Technion had, and still has, faculties, where the students choose the course of their professional development within the faculty. For instance, the mechanical engineering faculty offered specializations in energy, production, and other areas. Aeronautics, however, argued that the uniqueness of the field is its multidisciplinary nature, an airplane being an interdisciplinary and multidisciplinary craft. An aeronautical engineer is a better engineer, if he sees the whole picture, rather than just the aerodynamics, structure or control systems.

I, for instance, specialized in structures in my second and third degree studies, and only years later returned to the study of aerodynamics. I would not have taken that path if, during my first degree studies, I had specialized only in structures, without first being acquainted with the other areas. It is this exposure that creates within the student the openness to learning new things.

There are considerably fewer aeronautical engineers than there are mechanical or electrical engineers, and still, people ask me 'how come there are so few of you, and yet we see you everywhere we go?' The answer is that aeronautical engineers receive a multidisciplinary education, and are unafraid to enter unfamiliar territories later in their careers."

As aforesaid, this approach stems from the unique nature of an aircraft.

Aviv Rosen: "Airborne vehicles are technological systems that undergo optimization even as early as at the planning stages. They would not have been able to fly, otherwise. To illustrate this point, we can use an analogy with another discipline, such as construction. A construction engineer builds the structure; after him, the plumber comes and performs his work, and even if the two never coordinate, things will work out – the building will stand. Even if more weight is added or a groove is made, it will not make much of a difference. In comparison, on an aircraft, all the activity is coordinated and integrated, from beginning to end."

Systems Engineering Training in Israel

Aviv Rosen: "During the 90s, Israeli industry badly needed systems engineers, mainly in the technologically and systemically complex defense and aviation industries. The people in those industries felt that systems engineers who acted as such on their own accord were not enough; they needed to be given tools and methodologies to help them bring order to their applications of systems engineering. The recognition of the importance of systems engineering had already set-in abroad, mostly in the United States, and the Israeli industry wished to adopt similar patterns. Companies like IAI and Rafael began training systems engineers on their own, using internal training frameworks.

In the mid-nineties, representatives of these companies approached the Technion and asked it to establish an academic program that would grant its graduates an academic degree in systems engineering. The special affinity between aeronautics and systems engineering (and the fact that the request came from the aviation industries) led the petitioners to the then dean of the aerospace faculty, Prof. Aviv Rosen."

Why were they unsatisfied with their own, internal training programs?

Prof. Aviv Rosen: "It was a trend that started abroad, when more and more large systems failed. They were either technologically or economically unsuccessful, or simply failed to meet their deadlines. People saw that things were not working and found the problem to be *in the connections*. Consumers, such as the US Air Force, were the ones who put their finger on the problem: a lot of the time, the industry 'did not care' – it employed the 'cost plus' strategy.

The industry in Israel was not free of irregularities and mismatches either. One particularly notable phenomenon was the unexpected rise of new needs, which required changes to be made in mid-project. Technologically, the industry's output was impressive, but there was a sense of something not being right.

With the increasing academization of systems engineering overseas, the Israeli industries also wished to adopt the trend. This was not due to a lack of knowledge (they had very serious training programs that included hundreds of hours of study), but because they wanted an academic program that provided its graduates with 'qualifications' in the new field; a respectable program that added a different point of view and contributed to the field's advancement."

According to him, *systems engineering is a profession, meant, first and foremost, to serve the needs of the industry.* A profession born mostly out of practical needs in the field, not an area of study that the academicians wished to explore of their own volition.

What, then, was the Technion's interest in developing such a unique program, and agreeing to the special format of including industry representatives in the structuring and direction of the program, no less? After all, the academic steering committee was based on more than just academicians, and the approval of such bold initiatives is not an everyday occurrence in a respectable academic institution, the natural inclination of which, as these establishments go, is toward conservatism.

Prof. Aviv Rosen explains that parallel institutions abroad have similar attitudes: "The great impact of engineering institutions like MIT resides in the training of engineers. The training happens in research-oriented academic institutions, because an institution that performs research can train engineers better than one that occupies itself only with training."

And, in the context of Israel and the Technion, he adds: "The founding fathers of the Technion had established that one of its goals was to contribute to the needs of the State of Israel; training a systems engineer, who would benefit Israeli industry, is one way of realizing this vision."

The Technion's program is, as aforesaid, meant to train people from the industry, who have some experience in engineering work. Applicants are required to have at least three years of hands-on experience. The framework of the studies is also unusual, as it takes into account the limitations of students who are older than average and usually have families and work full-time. The students have one, concentrated weekly day of studies. Spread across two and a half years, the program's curriculum includes approximately 500 hours of classes (and many more hours of work outside the lecture rooms), and it places a heavy emphasis on practical assignments, done in teams. This is, of course, to foster one of the most basic elements of systems engineering – teamwork. It is also why the final project of the systems engineering training program, which, in a regular academic framework, would tend to be personal, is a group project (of course, each group member is graded separately).

The first class of the Technion's training program graduated in 2001. As of 2012, the program has trained approximately 900 systems engineers.

On the students:

At first, students who entered the program came mostly from the companies that had taken part in the initiative of founding it: Rafael, IAI, and the IDF's technological forces.

In the years that followed, the circle of organizations and industries that provided the program with students slowly expanded.

Prof. Aviv Rosen: "Today, those who come to study here are mostly mechanical engineers, aeronautical engineers, and electronics engineers, but some *computer science graduates* can be found among them as well. The education of the latter is, in most cases, focused almost exclusively on the world of computing, and they are not always interested in expanding their knowledge, because the market has a demand for them without it as well. But a software engineer who can 'speak other languages' has an advantage over software engineers who focus only on software. Software and systems engineering share many similarities, because computers are the glue that holds the various systems' components together (see also interview with Sharon Shoshany Tavory).

We also have some *physicists* among our students. Physicists make good systems engineers, because they have very good simplification skills, which they acquire thanks to the nature of physics as a discipline. The ability to simplify often allows a better understanding of complex systems."

Further Insights on Systems Engineering

Systems Engineering and Control

- Many systems engineers start out in the control field. This is due to the nature of the profession. Control is, by definition, integration, and so, it follows systems engineering principles. In aeronautics, control is often more systemic than in other fields.

Systems Engineering and Professionalization

- Systems engineering tends toward a broader perspective, rather than toward specialization. Engineering needs both specialized experts and people who can and want to see things from a wider angle. The "want" part is important, because not everyone is willing to give up specialization. Many people are very comfortable as experts; they enjoy being perceived as such and having people come to seek their advice. There are, however, others, who get "bored" after a few years of specialization and begin searching for something new. Engineers receive a lot of systemic stimuli. If they focus on specialization, they will remain specialized. But, if they have an open mind, a broad perspective, and instinctive leadership ability, they will naturally tend toward systems engineering.

On the Integration of Systems Engineering Culture in Organizations

- Changing an organizational culture is a difficult task, even if the CEO is convinced that systems engineering is very important. I asked one such CEO: 'If you have examples from past projects that prove how very beneficial systems engineering is, why don't you decree that no project in your organization goes without systems engineering?' The answer I received was: 'It doesn't work like that. If our man is not convinced, every time there is a glitch, he will tell me (the CEO): 'This is because you dropped all this systems engineering on me. It disrupts things, instead of helping.' People need to be convinced that the change is worthwhile. This is one of the advantages of our academic program: there are 900 graduates out there, in the field, right now, talking about it. And when they see an organization without systems engineering, they ask: 'Why no systems engineering?' This has an impact, but it will take time.

On the Qualities of a Systems Engineer

- If one is not personally inclined toward systems thinking, he will not be a good systems engineer. This trait is more essential than even training or experience.
- The process of developing a system (the V model, see page ... end of part one) often begins with the creative 'disorganized' systems engineer. As it progresses, it requires a mind less 'wild' and more administrative. If, during the execution stage, as the project ascends the V model graph, creativity levels climb too high, the project will become problematic. The ideal systems engineer changes with each stage of a project's advancement. The ability to put the right systems engineer in the right place makes the difference between failure and success. (See chapters on The Lavi and The Iron Dome projects).
- I am not a systems engineer, because, as I see it, a systems engineer is one who actively practices the discipline in the industry, while my career is in academia.

A Systems Engineer's Work Method

- A good systems engineer needs to know how to ask questions. He needs to tell the specialist underneath him: "I am not the expert, but your explanation is not convincing. Come up with a better founded explanation, or I will keep asking questions."

3.5

SYSTEMS ENGINEERING IN THE WORLD OF TRAINING AND CONSULTING

3.5.1 "COMBINING ENGINEERING AND MANAGEMENT SKILLS"

An Interview with Dr. Eric Honour

Systems engineering is a new discipline, created by the need for it, which originated in the field. As part of its evolution process, it is gradually establishing itself as a profession and an area of academic research and study. We talked with an experienced systems engineer, who has lived through this field's evolution alongside that of his own professional career, which included field work as well as teaching and research over the years.

Background

After graduating high school, Eric Honour joined the US Naval Academy, where he obtained a bachelor's degree in systems engineering.

Eric Honour: "Back then, during the 1960s, we had no courses in systems engineering processes, like they do today. Systems engineering was mostly about control system theory. Our studies focused on system analysis, a combination of mechanical

Managing and Engineering Complex Technological Systems, First Edition.
Avigdor Zonnenshain and Shuki Stauber.

and electronic systems. The software field was making its first steps at the time, and computer studies were included in the systems engineering curriculum. We did a lot of experiments and mathematical calculations. We received a combination of officers' and system analyzers' training."

He was discharged from the military 9 years later, in 1978, having served as a pilot in the navy for a large part of those years. During the last years of his service, he served as an instructor on engineering fields, like electronics and thermodynamics.

This basic training is unusual for a systems engineer. Most systems engineers are trained in one of the basic engineering fields and only then, having acquired some hands-on experience and having recognized their tendency towards systematic work, become systems engineers. Eric Honour's basic training, however, was in systems engineering from the start. This is because he began his career path in the US navy, an organization with unique characteristics and needs. This experience proved advantageous for him later on.

Eric Honour: "The fact that I had a military career before I started practicing systems engineering had two advantages for me. The first: it allowed me to acquire leadership skills, as an officer. Most engineers don't get that kind of training. The second: I had acquired knowledge about the operative use of complex systems, so that when I designed systems later on, I kept thinking about how people would use them. People purchase the system to use it in combination with the other systems in the operative environment. I was able to look at the systems not only as a developer and designer, but also as a user."

(See interview with Henry Broodney for a discussion on a similar topic.)

(In this context, see also the discussion with Hillary Sillitto about the mobile phone becoming the number one cause of car accidents.)

After completing his military service, he spent a few months working in software development and then joined E-Systems, a company engaged in military communications systems, where he worked until the late 1980s. During this time, he had served five different positions, most of them in system engineering on various organizational levels.

In his first job as a principal engineer, he designed electronic circuits for signal processing systems. Afterwards, he was promoted to an administrative position as the systems engineer in charge of system design, where he had to manage the engineers who performed the detailed design of the systems' components. With a job title of Engineering Supervisor, he managed an engineering team of nine. His job was to coordinate between the different factors in the design of systems for various projects in the communication intelligence field – military communication systems that allowed one to listen to enemy communication traffic.

He continued to advance, and for the next 3 years, he served as an engineering manager with nearly 50 engineers in three groups under his authority. In this position, he also continued to be actively involved in the systems engineering of larger systems worked by those groups, including being the designated lead systems engineer for two major systems. After this experience, he was offered to leave the systems engineering and take up project management.

Eric Honour: "I began working with costs, schedules, stakeholders' involvement, preparing proposals for clients. I did this for 3 years, but after all that, I chose to return to systems engineering, because I realized that I loved the professional, technical work more than administrative work. During my time at E-Systems, I got to work on projects at all stages, from requirement formulation to delivery."

Being well aware of what he wanted to do, Eric Honour began working for Harris Corporation, a company that also specialized in military communications systems, while also working on major systems for NASA and for the FBI, where he spent another 10 years, this time, in only one position: Senior Principal Systems Engineer. As part of this job, he was integrated into various projects within his area of expertise, including projects in communications intelligence, smart guidance systems and more.

In the late 1990s, the company encountered some economic difficulties and was forced to make company-wide pay cuts. This circumstance led Eric Honour to his decision to resign and establish his own business, a company that offers consulting and training in systems engineering fields.

Eric Honour: "I realized that job security in large companies was no longer guaranteed. I was also the president of INCOSE at the time; I had an international reputation and good connections, so the timing was good. At first, I had aimed toward consulting only, but over the years it became evident that training courses were better business."

The company employs seven experienced instructors who lead dozens of focused courses a year, mostly on client-site. It also provides consulting and research services. Recently, Eric Honour has also completed his dissertation and received his PhD.

Professional Background and Systems Engineer Training

Eric Honour believes that the basic engineering disciplines that provide a good knowledge infrastructure for a systems engineer are electronics engineering, mechanical engineering, and software engineering; because most systems include these three core fields within them.

He is not among those who denounce basic systems engineering studies.

Eric Honour: "Today, the convention regarding systems engineering training is to study for a master's degree in systems engineering, having already gained some experience; but there are undergraduate programs available in the United States, and I believe it is entirely possible to study this field without prior experience. After graduation, one can gradually be integrated into the practical field. It is not very different from an electronics engineer who studies for 4 years and then starts working at an organization, having no prior experience. Clearly, his first assignment will not be designing a complex electronic system. He will first be attached to a mentor and charged with a relatively simple task within the framework of designing a complex electronic system. Similarly, a newly graduated systems engineer can first be assigned less complex tasks and aided by a mentor. Much of the knowledge I acquired in this field was gained by experience and through mentoring."

Eric Honour, owner of a company that provides systems engineering courses, explains the difference between the activity of universities and companies like his in the systems engineering field: "Most of the universities' activity in the field consists of long education programs that grant academic degrees to those who complete them. Many of the lecturers in these programs are academics with no hands-on field experience. There is no competition between these studies and the ones provided by training companies; the two complement each other. However, in recent years, in order to generate more income, universities have begun to offer on client-site short-term courses. For this purpose, they also offer instructors with field experience. This particular activity indeed competes with private training companies."

The question, then, is: if in the past, the added value these companies possessed was lecturers with vast experience, which academics could not offer, what added value can these companies offer their clients today?

Eric Honour: "Because of their structure, universities can't always pay good, experienced lecturers salaries worthy of their status and are, therefore, often unable to employ the best ones. Private training companies do not suffer from this limitation, being able to bring the best lecturers and customize short and focused courses to suit the needs of a specific company's employees. It's not just about teaching processes or administration, it's about thinking systematically. It's not easy to teach, and so, training has to be left to those who are experienced and able to bring good examples from their personal experience."

It is his belief that there is a fundamental mistake in the way systems engineers are trained today. According to him, too much emphasis is placed on processes, rather than instilling system analysis skills: "That is how systems engineers become process engineers. Instead of telling others what to do, they do things themselves. This is wrong. It is impossible to construct a system without system analysis. That is the systems engineer's job, because none of the core engineering disciplines include the knowledge of how to perform a system analysis."

He says that a systems engineer needs to be trained as an engineer, but also needs knowledge in administration and project management, namely, the management of tasks and schedules.

He also stresses the importance of leadership: "A good school will teach more than just systems engineering courses, but project management courses as well. And if it's a really good school, it will also teach leadership, the ability to motivate others, to make a group work together. Leadership may be hard to teach, but some leadership skills can be taught, despite it being an inborn trait."

Eric Honour divided systems engineer training into three areas: technical, administrative, and leadership. In his opinion, both the technical and the administrative areas can be taught in courses and through implementation in the field – by gaining experience at work. However, the best way to instill leadership skills and teach one to think like a leader is mentorship combined with experiential teamwork. He says that solving problems together helps build teamwork skills and helps the students learn how to work with other people.

He testifies that he had learned a lot from a mentor who had accompanied him at E-systems: "Mentorship is a powerful training tool, and it's a pity companies don't use it more widely."

The Evolution of Systems Engineering

Eric Honour says that 40 years ago, when he had taken his first steps as a systems engineer, systems engineering had been radically different from what we know today. During those years, it had undergone some significant changes, yet the basic patterns designed in the 1960s were similar to those used today: "In the late 50s, a book was published, that laid the foundation for systems engineering. About a decade later, in 1969, the first systems engineering standard was published, that set standards and presented processes similar to those we see today. When I studied at the academy, the emphasis in systems engineering was more on systems analysis, rather than processes and standards. In the 50s and 60s, systems engineers wrote textbooks and standards for what was rapidly becoming a professional field based on analysis and common sense – the right way to do the job. By the early 70s systems engineering had full structure as an engineering discipline."

However, according to Honour, the SE discipline largely fell into disuse through the 1970s and 1980s, except in some narrow areas. What caused this change?

Eric Honour: "In the early 1970s, managers were viewing systems engineering as 'just common sense,' and they began to put less emphasis on it in projects. Funding for SE was reduced. There were some areas where it continued strong, such as the Navy's Aegis (Naval combat systems) program and the NASA space shuttle development. During those years, these organizations were deeply engaged in systems engineering, but the field did not evolve outside of them. Only during the 1980s, after the breakthrough in the software engineering field, did awareness of systems engineering begin to grow again. This change was due to a significant problem encountered in the late 1980s: many software failures were discovered, because software personnel had not received the information they needed, at the quality they required. They began looking at ways of receiving better requirement specifications, and that put systems engineering back into people's minds."

However, many claim that software specialists only care about software.

Eric Honour: "True. And so, to write better software, they wanted better requirements specifications. In time, their approach changed. In the last 20 years, they switched from an approach based on the desire to receive better requirements to one relying on client relations, striving to structure the requirements together with the client."

He believes the evolution of systems engineering also depends on it becoming a scientific discipline: "Take electronics for example. In mid 19th century, people built electric engines. They had no idea how they worked, they just knew that they did. It had taken many years for the link between electricity and magnetic fields to be discovered and mathematical tools to be formulated, allowing the mechanisms of generators and engines to be analyzed. These discoveries created a new world of electronics, radio and frequencies.

So if we want systems engineering to get better, we need scientific research that includes three stages of development: analyzing events that show us what systems engineering is (your book is a good example of this), empirical research and, finally, theoretical research. Today, systems engineering is somewhere between the first and second phases."

More Insights on Systems Engineering

On the essence of systems engineering

– "Systems engineering is several things: it is a discipline, because it articulates a list of processes that, if implemented, provide us with better results. It is also a way of thinking, because if we don't think correctly, the processes by themselves will not be much help to us, because the core of systems engineering is systematic thinking – the ability to see the big pictures and base decisions on it. It is also a profession; because there is a group of people who have embraced this way of thinking and these processes in order to use them to do things better. When people ask me what systems engineering is, I explain that each part in an airplane is made by a different engineer, while a systems engineer looks at how all the parts come together. He tells each engineer what to design within his specialization. This means systems engineering combines engineering and management. A systems engineer needs to possess both administrative and engineering skills. He can't tell the engineers what to design, if he doesn't know how it fits into the system; and he can't tell them what to design, if he doesn't have good relations with them."

On systems engineering and project management

– "Systems engineering and project management differ when it comes to priorities: project managers focus on the tasks, schedule and budget. The technical manager (or chief systems engineer, or whatever you call him) is responsible for the results the tasks yield.
The project manager wants to accomplish the mission. The systems engineer wants to make sure it is accomplished well. Their goals are the same, but their priorities are different. The project manager looks at the cost first, then at the schedule and only then at the technical aspects. The system's engineer's priority list is reversed."

Eric Honour demonstrates these differences

– "Let's say a problem is discovered in the link between software and hardware. The principal systems engineer comes to the project manager and says 'there's a problem here, because the software engineer and the mechanical engineer have chosen different paths, and now we need to perform a forward analysis and decide where we go from here.' The project manager then asks how much this will cost, and the principal systems engineer says 'fifty thousand dollars, because we need to make a software simulation to assess the gaps.' And then, the project manager says he doesn't have that much money."

On systems engineering and intercultural differences

- Eric Honour finds that intercultural differences are mostly differences in language and the different conceptualizations adopted by systems engineers in different countries, or even different industries. However, the systems engineering thought patterns, the ways we look at the big picture, are fairly similar.
- He notes a relevant, significant difference between the United States, a culture that pushes forward to get results, and European culture, prevalent in countries like the United Kingdom and Germany, where the cooperative approach is more common. He also mentions Israel and its high motivation and entrepreneurial passion.

On a systems engineer's skills

- People choose to study engineering because they love things more than they love people. So, at first, as engineers, we learn about things. Then people grow to become systems engineers, because they realize that it's impossible to build complex systems without understanding people as well as things. They understand that it's more important than mechanical or electronic or software design.

3.5.2 "MODEL-BASED SYSTEMS ENGINEERING"

An Interview with Sanford (Sandy) Friedenthal

Systems engineering is an evolving discipline that requires one to possess, among other things, hands-on experience and training in many facets of engineering in order to be applied effectively. In this chapter, we describe how an engineer became a systems engineer in US aerospace and defense industry, and we discuss the path he took to eventually become an expert consultant in this field.

Background

Hughes Aircraft Company:

After graduating from a general engineering studies program at UCLA in 1973, Sandy Friedenthal began working for the Hughes Aircraft Company, Missile Systems Division in Canoga Park, California. Hughes Aircraft was a major US aerospace and defense contractor that was owned by the Howard Hughes Medical Institute, a non-profit medical research organization founded by Howard Hughes. The Missile Systems Division developed and produced many state-of-the-art missiles including Maverick, Phoenix, and TOW. The company maintained a culture of technical excellence and a long-term view to advance critical technologies.

Sandy joined the missile guidance lab, where the focus was on developing and applying advanced missile seeker and guidance technologies to missile system design. Sandy gained hands on-experience working in a high-technology engineering lab and was mentored by technical experts in the field. The work included development of prototypes using radio-frequency traveling wave tubes, electro-optics sensors, and guidance algorithms.

After a few years working for Hughes, Sandy decided to continue his education towards a master's degree in control systems.

Sandy: "I chose the control field based on personal interest and practical considerations. The field of study was highly relevant to my work at Hughes, but had an analytical orientation that was of interest to me. As I found out later, the control systems theory turned out to provide an excellent foundation for systems engineering."

In his job, he was able to apply the analytical techniques he learned in school. The application of control systems techniques was all about modeling the system and analyzing its performance. However, the work remained focused on the control aspects of the system with limited emphasis on other multi-disciplinary aspects of the system. He was using systems engineering methods, but only within the framework of my specific discipline.

The company's organizational unit included specialized engineering departments that focused on different aspects of the missile, such as guidance and control, propulsion, aerodynamics, and warheads. "From my perspective as a young engineer, there was not a lot of cross-disciplinary integration. The integration occurred at the program level by special teams in charge of integration and test."

Martin Marietta Corporation, the Early Days

Sandy left Hughes Aircraft in 1981 for a new opportunity at Martin Marietta in Orlando, Florida. Sandy had been exposed to Martin Marietta as part of a proposal effort at Hughes Aircraft, and was impressed with their proposal contribution.

Martin Marietta was a large and diversified US aerospace and defense contractor. Sandy joined the Advanced Concepts Group in their Electronics and Missile Systems Division, where he worked on new concepts for missile systems. He had the opportunity to work on an advanced concept for an air-to-air missile that integrated technologies across the entire missile system including seekers, guidance and control, propulsion, airframes, and warheads. The advanced concepts projects were not just on paper, but involved conceiving, developing, building, and testing the concepts.

Sandy found the culture of Martin Marietta quite different from that of Hughes Aircraft: "Not having to focus on profit, Hughes Aircraft maintained a long term time horizon. They often planned their research programs out 10 years. In contrast, Martin Marietta, which was a for profit company, planned 3 years ahead. The difference between a 3 year time horizon and a 10 year time horizon had a significant impact on the pace of work. (Today, planning even 3 years ahead is considered long term). There was a constant sense of urgency; where schedule mattered, even within the Advanced Concepts Group.

Additionally, unlike Hughes Aircraft, where the work was performed as part of the specialized departments, Martin Marietta had a matrix organization where people were assigned from their specialized functional department to work on a project with other technical and non-technical disciplines. In addition, at Martin Marietta, the manufacturing plant and the engineering development were located in the same place, unlike Hughes Aircraft, where the engineering and manufacturing were performed in different states. As a result of these business differences, the nature of my work was significantly different at Martin Marietta. I enjoyed the pace at Martin Marietta, and the exposure to other disciplines and programs, which helped me gain a broader perspective."

Sandy emphasizes: "This environment created partnerships between disciplines working on the general design of systems. Systems engineering methodologies, as well as the term itself, were used at Martin Marietta. I was able to support overall systems development on a daily basis."

Evolving as a Systems Engineer at Martin Marietta

Sandy transitioned from the Advanced Concepts Group to work as a systems engineer on a complex electro-optical system that was transitioning to manufacturing. There, he helped to support engineering changes, and to document and improve the way the system was built and tested on the production line. Over the next several years at Martin Marietta, he had increasing levels of responsibility in systems engineering. In the mid-1980s, he became manager of the Systems Requirements, Design, and Integration Section, and later became Director of the Systems Engineering Department.

At the request of the VP of Engineering, Sandy then took the responsibility to lead a new company initiative in concurrent engineering. Concurrent engineering is an approach that is fundamental to good systems engineering. It is all about involving the right disciplines early enough in the design process, so that down-stream processes such as manufacturing and support are adequately considered. Engaging these disciplines early in design fundamentally changes how the work is done, and the methods, tools, and training have to be adapted accordingly.

Sandy Friedenthal: "We formed a work group of senior-level executives from different disciplines, and together, we formulated a strategy for how to implement concurrent engineering across the organization and on projects. We also worked with our customers to involve them in this strategy. It was an all-encompassing initiative with many facets: from planning, development of the methods and tools, training personnel, and working with projects to implement the approach. The application of concurrent engineering to the Navigation and Targeting System on a particular advanced helicopter project was widely publicized across Martin Marietta and their customer community, and was considered a successful early adopter of this approach. In the late 80s, Martin Marietta acquired another company's operations that developed reconnaissance systems. I joined the program as the technical director (one of the names organizations use for lead systems engineers. See, for instance, the interview with Mimi Timnat from Elbit Systems). Much of the work was subcontracted out. I applied systems engineering to the management of the subcontractors and to ensure the overall system requirements were satisfied. There were also many organizational challenges, due to mix of cultures between Martin Marietta and the acquired company."

Continuing Career at Lockheed Martin

Martin Marietta merged with Lockheed to become Lockheed Martin in the mid 1990s. Sandy and his family moved to the Northern Virginia area to continue his career with Lockheed Martin. There, he had many opportunities to grow as a systems engineer and work in a variety of technical and management positions. Sandy performed in many different systems engineering roles, some on projects and others leading engineering improvement initiatives.

In his last years at Lockheed Martin, Sandy led up the Corporate Initiative on model-based systems development (MBSD), where he was responsible for developing strategies to deploy a model-based approach across the Business Units, and to provide direct MBSD support to the programs. MBSD, or model-based systems engineering as it is more generally known across industry, involves formalizing how systems engineering is performed through the use of system models. Sandy left Lockheed Martin after a long and interesting career at the end of 2010, and became an independent consultant in model-based systems engineering

Sandy did not begin his career as a systems engineer. Rather, he evolved to become a systems engineer, and it became his job title later. The systems engineering role continues to evolve and broaden throughout his career.

Based on his personal experience, Sandy explained to us how he evolved to become a systems engineer. "You don't typically start off as a systems engineer. For me, it was an evolution where I started with a particular technical focus in the lab primarily dealing with electronics, then my technical focus shifted to guidance and controls, followed by increasing involvement with other technical disciplines, and different phases of a system's development from conceptual design through manufacturing. As I moved through my career, I had increasing exposure to other aspects of the system. The challenge for how these different aspects of the system and the project work together continued to present itself. I was intrigued by this challenge."

Sandy continues: "My background in control systems gave me a starting perspective for how to think about systems. My exposure to multiple disciplines enabled me to gain a diverse perspective on the nature of systems. Since I had the opportunity to support a variety of programs, I also began to see how systems engineering could be applied to different systems."

Sandy found that the principles of systems engineering, could be applied to things other than technological systems. Many of Sandy's assignments were associated with leading organizational initiatives. The systems approach was extremely useful approach to deal with large complex organizational challenges, where many stakeholders are involved with widely varying needs. The systems approach can be applied to help create some order from the chaos. As the lead for the MBSD initiative at Lockheed Martin, Sandy found it useful to think of the overall modeling infrastructure of modeling methods, tools, training, and project support as a system that has its own life cycle and interfaces. The systems approach enabled him to deal with the various complexities and aspects of the modeling infrastructure in a holistic manner.

More Insights on Systems Engineering

On the evolution of a systems engineer

- "Formally, I became a systems engineer when I joined the Systems Requirements, Design and Integration Section at Martin Marietta (during the first half of the 80s, see the preceding text). The systems engineers recruited into the department came from various fields of engineering. Back then, there was no one prominent field that produced more systems engineers. Today, we are seeing more software engineers and architects transition to systems engineering. This is in part due to the increasing scope of software in systems.
  Eventually, what leads people to systems engineering is, for the most part, their personalities. Analytical skills, teamwork ability, and a passion to understand how the pieces work together are of great importance.
  It is possible to learn how to be a systems engineer. It is a continuous learning process. Every time I work with a new discipline, I try to examine how I can adopt their perspective into my thinking about systems. In order to succeed in this, you need to establish some framework for thinking about systems."

On System Thinking

- "Today, when I work with systems engineers on a project using model-based systems engineering, I try to employ classic systems engineering methods: requirements, architecture, trade-off, the 'illities' (a generic suffix used for various system abilities; see also interview with Olivier de Weck) and verification. However, I also try to apply system thinking in everything I do, as a way of approaching problems. The focus of this approach is to understand different stakeholder perspectives and concerns, and define a problem first before jumping to a solution. Then establish value from the perspective of the stakeholders, determine alternative approaches to address the problem, evaluate the alternative solutions, and validate the solution addresses the need. System thinking provides a way to think about how the pieces of the solution fit together to address the problem."

Opportunities and Challenges

- "One of the difficulties in learning and communicating the benefit of systems engineering is the abstract nature of what systems engineering produces. Mechanical engineers focus on the mechanical aspects of a system and produce mechanical designs. Likewise, electrical engineers focus on the electrical aspects of a system, and produce electrical designs. The products these disciplines produce are very tangible, such as a computer aided design of a mechanical assembly or a circuit card. Software engineers also produce a tangible product, namely, code. On the other hand, a primary product of systems engineering is the specifications of the system components and the supporting analysis and data that demonstrate how these components integrate to accomplish the system objectives. The systems engineering products are more abstract than the products produced by many other engineering disciplines.

A critical characteristic of a systems engineer is the ability to think abstractly. This is a challenge for many people who either do not want to or are unable to think this way. A systems engineer must strive to see the picture in its entirety."

Sandy concludes: "I believe the more widespread adoption of systems engineering and systems thinking can provide an important contribution to help solve current and future problems. However, based on my experience, it takes a long time to build an effective framework to think about systems as a whole. We see evidence every day in systems development, in business, and in government, where considerations of the whole are compromised at the expense of considerations for particular pieces or facets of the whole. Perhaps what is needed is to introduce systems thinking beginning in early education, and then reinforce this framework throughout the educational process."

3.5.3 "THE MAIN REQUIREMENT: KEEPING UP WITH SCHEDULES"

An Interview with Niels Malotaux

One of the main problems that arise during projects is failure to meet deadlines. Consequently, a large number of projects are not completed on time. So, a large number, in fact, that many, including quite a few systems engineers, believe such situations to be acts of fate, often brought on by developments that cannot be foreseen. This makes it a self-fulfilling prophecy.

Niels Malotaux, working as a Project Coach, challenges this perception, because his experience tells him that projects do not have to be late. Over the years, he has developed the techniques to put this into effect.

This chapter will deal with Malotaux's perception of systems engineering, emphasizing the issue of failing meeting deadlines and what to do about it.

Background

As an electronics engineer, Niels Malotaux devoted the first half of his career to the development of systems with integrated hardware, electronics, and software. For 18 years, he has been managing an electronic systems design company in the Netherlands, alongside his brother, followed by 15 years of Project Coaching, developing, and honing techniques for successfully running projects in real practice.

Niels Malotaux: "As electronics engineers we specialized in systems design. When farmers approached us with the need to develop a system for climate control of pig-stables, feeding the pigs, as well for example weighing chickens, we looked at the whole system, including the interfaces, the sensors, and actuators, the PC at the farmer's bedside table, how it would be used, and how it could easily be programmed by local software people without knowledge of the details of the electronics hardware."

Niels Malotaux does not view himself as a systems engineer: "Systems Engineers accuse 'normal engineers' of 'silo-thinking,' not minding if their developments don't work with other parts of a system. Now, what's the point of completing part of a system if it doesn't work together with the other parts? So, any engineer must at least take into account how his development will work together with the other parts. When I first heard about systems engineering I was surprised. I thought that was something every engineer should do, so why would we need the extra word? Therefore I wouldn't call myself a systems engineer, but I think I do meet the definition of one.

Engineers must be able to communicate with one another. First, they must be experts in their own fields, but at the same time, they should be able to communicate with people in other disciplines, to make sure the system as a whole works as it should. This is what I have been doing all my life and now I'm teaching systems engineers how to do the right things at the right time. In short: 'Quality on Time'."

Malotaux's words suggest that this issue receives insufficient attention in basic engineering studies. This is one of the reasons for the rise in the importance of systems engineering, a discipline meant to bridge over gaps.

One of the fundamental practices in systems engineering is defining a system's requirements. According to Niels Malotaux, a systems engineer has to understand the client's *real* requirements. In many cases, clients do not know how to define their needs correctly, leaving it to a systems engineer to figure them out: "We used to design electronic products for our clients, but it was not enough to merely do what was asked of us. For example, working with a client I asked him what his problem was. He answered: 'Just do as I tell you,' to which I said: 'But what is the problem?' He said: 'What problem? I don't have a problem'. So I said: 'If you don't have a problem, we don't need to do anything'. 'Ah. Well, the problem I want to solve is ... ' This is an example of the 'ask five times *Why*' technique and an example of my work method, as a systems engineer. First and foremost, I look for the client's real need."

In fact, we too have witnessed this approach during our conversation with Malotaux. When asked certain questions, he enquired as to what we really wanted to know. For instance, when we asked when he had first encountered the term 'systems engineering,' he wondered why we were asking. Having learned that our purpose was to discern the root of the term, he proceeded to tell us his version of it.

The Essence of Systems Engineering

When Niels Malotaux told his father he was planning to participate in a systems engineering symposium, his father asked him what that was. Niels told him: "Systems

Engineering is what you lectured your students on at University. You only didn't use the term Systems Engineering. Only you taught them much more than what Systems Engineering curricula teach nowadays."

Niels Malotaux: "Other than we see at other universities, my father didn't want to create specific Systems Engineers. He argued that every engineer should know about Systems Engineering, how organizations work, how R&D works. While discussing about the whole system with fellow engineers, they still have to be able to drill down to their specialized field to be able to see and communicate the consequences for their field. Optimizing the total system, which may mean suboptimizing one's own contribution to the system.

Systems and *engineering* are just words, and the problem with words is people attribute different meanings to them. In Dutch, an engineer is someone with a university degree in engineering, where I feel that in English an engineer is anyone who's using a screwdriver professionally. Systems engineering is everything that makes a system successful. The word *engineering* can be attributed to the act of organizing things in a way that makes them work. This can include anything. The name systems engineering is simply the label given to all the things systems engineers must do in order to create a successful system. It is much like when people talk about terms like *lean*: sometimes these words help us talk about our fields. This is why I accept people talking about systems engineering. If it helps, it doesn't matter what it's called, but using a common term helps to communicate better."

As he sees it, systems engineering is an attitude that can be adopted by anyone, but, in effect, focuses on technological systems, which is why a systems engineer must first be an engineer. Still, a lot of engineering as well as systems engineering knowledge can be used to organize social systems.

He adds: "I teach systems engineers that when working on a project, failure is not an option, because the knowledge of how to routinely make projects, even very complicated projects, successful and on time is known, although apparently not too much practiced. As soon as we see that things can go wrong, we don't call it fate (as in Murphy's Law). We do something about it before it goes wrong."

In the late 1990s, when Niels Malotaux first heard the term *systems engineering* ("people invent the term and make a hype"), he found that the newly defined discipline lacked sufficient attention to two important areas. The first of these was awareness of schedules, mostly concerning on-time delivery, on which we shall extrapolate later.

The second area was the human factor. It took quite a few years for systems engineering to understand the importance of the human factor, which only recently began to receive the attention it deserves.

Niels Malotaux: "People are always part of a system. A system that doesn't include people is a useless one. A systems engineer needs to understand how people behave: not how he thinks they should behave, but how they really behave – the two are not the same. In projects we have to think about the behavior of people working in the project, as well as people who will be using system developed.

They say people are unpredictable because they never do what you expect, and maybe, to a certain degree, that is true. But, for the most part, people are a lot more predictable than we realize. What we call 'unpredictable' is often simply different from what we expect. It is therefore very important for systems engineers to learn how people *really* behave. If we model the users of our systems according to what we *hope* they do, the system will not work."

He demonstrates: "Our company developed a monitoring system for controlling the freezing and defrosting of dough in bakeries. Historically, a baker had to get up at 3:00 in the morning to prepare the dough for the bread, to switch on the oven and put the carts with the dough into the oven. Now the dough has been prepared and frozen the day before, and the next morning it has to be slowly defrosted (if the dough is defrosted too quickly, it becomes wet and its quality suffers). We developed a system that performs the freezing and defrosting up to the point when the baking could begin. When the baker arrives in the morning, the oven is already switched on and all he has to do is drive the cart with the dough in the oven, and in half an hour's time, the bread is ready.

First we designed the user interface technically, with a menu-structure that allowed the baker to see the status and change settings systematically, we thought. However, when bakers saw this user interface they said: 'That's too complicated. We want to see the status at a glance from a distance and be able to operate the daily functions without going into menus.' So we redesigned the user interface accordingly. To see whether it was intuitive enough, every time we got a visitor in our office, we showed him the display and said 'Do something'. The visitor then said: 'What should I do?' We said: 'Whatever. Just try what you can do'. This way we learnt a lot of how to make the user interface as intuitive as possible.

To developers I usually say: 'We are already too much distorted. We cannot imagine what normal users find normal. So we have to find out by giving it to them and *observe*. We need the feedback from normal people.' "

An Insight on the Importance of Planning Ahead

Niels Malotaux: "Intuition is how we automatically react on situations. Intuition is fed by experience (if it's from before birth, we call it instinct). If intuition would be perfect, everything we do would be perfect. As not everything we do is perfect, apparently our intuition sometimes steers us in the wrong direction. So, instead of just doing things intuitively, we better first plan *what* to do and *how* to do it most efficiently. The plan must be doable and we have to do as planned. Once we did that we can check whether the result was according to the plan and whether the way we did it was as efficient as planned. If it wasn't, we think whether we should and how we would do it better. If it was, we think whether and how we would do it even better. Then we decide, based on this analysis, what to do differently the next time we plan and do. This way we continuously improve what we do, how we do it, as well as the processes we find to do it best. If we do this very quickly and frequently, we quickly learn how to do things better all the time. This is what Deming already taught with the Plan-Do-Check-Act cycle. It's a very powerful technique."

Meeting Deadlines

Many project leaders will testify that a project's three main objectives are to meet the specified requirements on time and on budget. However, if the time and the budget are fixed, the result is by definition not fixed, whether we like it or not. Once we realize that a lot of things we do in projects later turn out not to have been necessary, there is a lot of opportunity to keep what is really required within the time and hence within budget.

One of the main reasons for the evolvement of the systems engineering field is the fact that engineers mostly focus on achieving the first objective, namely, meeting the project's requirements and even only their own part of the requirements. Systems engineering is the bridge that leads to the two remaining objectives. Nonetheless, having emerged from the engineering world, a large part of the discipline still aims to address the technological complexity, rather than the last two objectives: meeting the budget and on-time delivery.

Niels Malotaux: "Delivery time often is one of the most important requirements of a project. How come most projects are late? Apparently all other requirements are treated as more important than the most important one. Weird.

The project manager is *responsible* for the right results at the right time. The systems engineer, however, *determines* the time and what it will cost. This makes the systems engineer (as well as all other workers in the project) at least as responsible for delivering Quality on Time."

Malotaux believes on time delivery is more important than staying on budget, because one brings about the other anyway: "One of the first things I ask when starting to coach a project is: 'What is the cost of one day of delay?' Usually people don't know. Then I ask: 'What is the cost of one day of the project?' Usually they don't know. Then I ask: 'What do you cost per day?' Most people don't know.

When discussing these questions in a project, I just saw, trough the small window of the door, the boss passing by. I opened the door and said to the boss: 'Boss, these people don't know their cost, the cost of the project nor the cost of one day of delay. How can these people make *design decisions*?' He said: 'I don't know their cost, but I'll find out!' An hour later he returned, saying: '€400 per person per day.'

The benefit of the project should be huge, otherwise we should do another project, but if we don't know the benefit I always suggest to assume that the benefit is about 10 times the cost of the project.

Using this figure we calculated the cost of one hour of delay: In this case it was 7 people × €400 × 10 per day. This is usually a lot more anybody had imagined and it is a good basis to start making much better founded design decisions.

Some engineers complain: 'We may be delaying, but this way the product will be even better!' My reply: 'What is the return you get for that improvement? Does it justify the added expense?'"

Believing that about 50% of the work done on a project later will prove not to have been necessary, there is a lot of opportunity to save a lot of time. Niels Malotaux uses several techniques to deliver the right things at the right time. For example, to improve our efficiency, he lets people plan every week: "Many people use checklists of what to do. However, hardly anyone checks whether what he thinks he has to do fits in the available time. So we determine how much time we have available in the coming week. Then we take about 1/3 of that for all the routine interrupts we get, like planning, meetings, email, telephone, drinking coffee, helping each other, etc. For some people these interrupts consume 100% of their time and hence they never have time for what they are supposed to do. Therefore we time-box it to 1/3 of the available time, leaving 2/3 for whatever we plan to do. We estimate the tasks we have to do and fill the available planed time exactly. Every hour we do not plan will evaporate. Every hour we plan to do more than we have time available, will not be done anyway, so don't waste time planning something that we already know won't be done.

On a spacecraft project with many sub-contractors, the engineers were complaining that they didn't get enough time to properly do what they had to do. In this project all engineers were trained systems engineers with a lot of experience, except for how to be on time. I asked a systems engineer why he was complaining. He said: 'I have so many documents to review before we have a meeting with the sub-contractors. We cannot do a proper reviews if we only get two weeks to do it. This causes a lot of stress'. I asked 'How many reviews?' 'Seven'. 'How much time for each?' 'I don't know'.

'Is it more than two hours and less than seven?' 'Well we have heavy and light documents'. After some more discussion we found that four heavy documents would take about 15 hours each and three light ones about 2. OK, that's $4 \times 15 + 3 \times 2$ is 66 hours. What else do you have to do? After some discussion we ended up at 99 hours of work to do, while the available time was 46 hours. At this time we knew that just starting and doing our best wouldn't be successful. As 'failure is not an option,' I suggested to organize the reviews more cleverly, and I gave them some suggestions how to do that. The result was that all documents were reviewed properly on time before the meeting with the sub-contractors. Everyone was satisfied.

Before having coached this project, they never met any deadline. It's now about a year later, and they haven't missed any deadline since. No tools, just some simple techniques."

A conclusion easily reached from Malotaux's words is that delays are not an unavoidable obstacle brought on by fate, but problems that can and should be resolved. "If in *retrospect* we see that we spent more time than necessary, the time is already lost. If in *prespect* we see that we are going to waste time, we still can refrain from wasting it. Isn't it strange that the word 'prespect,' while so important, doesn't exist in English?" First, however, the problems must be defined. The situation in the aforementioned example shows us just that. Niels uses more techniques, like bi-weekly deliveries to improve our effectiveness, and a TimeLine process to see what will happen and to do something about it if we don't like what we see.

Niels Malotaux: "I tell them to solve the problem as systems engineers, to treat the timely delivery problem as a design problem, and then use their expertise to solve it. In the example I gave, once the team became aware of how much time they had, what they believed they had to do (and whether they should do it) the problem was resolved, and the schedule was met. It is important to work one step at a time, to ask ourselves what needs to be done next and move on to planning the next phase."

INDEX

Managing and Engineering Complex Technological Systems, First Edition.
Avigdor Zonnenshain and Shuki Stauber.